Contents

Preface

I had little idea, when I was asked about shortcomings in available textbooks, that I would end up writing a book myself, but here it is and I hope that you find it useful. My immediate comment was that there was no textbook available for teaching mathematics to geologists which was pitched at the right level. There are some good texts but, generally, they are aimed at students who have a good grasp of mathematics and wish to apply their knowledge to geological problems. There are also good books which cover mathematics at a level more appropriate for the majority of students who did not follow mathematics to the end of their secondary schooling. Unfortunately, these books cover mathematics as a subject in its own right, rather than as a tool for application in other subject areas. There are no books specifically aimed at teaching basic mathematical techniques to geology students.

This is my attempt at such a book and my approach is to use geological examples to illustrate the meaning of the mathematics rather than to discuss how mathematics is applied to geology in practice. Having said that, I hope that many of the examples used in this book are useful as well as illustrative. I also hope that this book will encourage more geologists to use simple mathematical arguments when they are appropriate. There is a tendency for many otherwise well trained geologists to avoid mathematics, although application of a few simple mathematical methods would make their work both more convincing and more testable.

Many people have been encouraging and supportive during the writing of this book. First of all, Una-Jane Winfield from Chapman & Hall who originally persuaded me to write this and was always enthusiastic. Unfortunately, Una-Jane left Chapman & Hall before the project was complete and I should therefore also thank Dominic Recaldin who took over this book at Chapman & Hall at very short notice. The most important people of all have been the students to whom I have taught this material over the years and who have been responsible for its mutation from a simple abstract mathematics course into a more relevant geology-oriented course. In particular, the 'class of 92/93' who were guinea-pigs for this book and who spotted several short-comings. Other problems were spotted by Stuart Hardy and Chris Willacy, postgraduate students of mine at Royal Holloway, University of London. However, the biggest contribution to improving my original drafts was undoubtedly from the reviewers, Andy Swan at Kingston University and particularly Peter Gould at Liverpool University.

David Waltham
Royal Holloway,
University of London.

Mathematics as a tool for solving geological problems

1

1.1 INTRODUCTION

This book is not about specialized geological mathematics. Mostly, this book is about simple mathematics, the sort that many people are introduced to at school. However, such mathematics is frequently poorly understood by geology undergraduates and few students are able to use the maths they know for solving realistic problems. The objectives of this book are to improve understanding of simple mathematics through the use of geological examples and to improve the ability to apply mathematics to geological problems.

This is not a formal mathematics textbook. My aim is to try to instil an intuitive feel for maths. I believe that this is more helpful than a rigid, formal treatment since formality can often obscure the underlying simplicity of the ideas.

Although this book concentrates upon standard mathematical procedures, it does contain a few more specialized techniques. The majority of the mathematics encountered by typical undergraduate students is therefore covered here. The exception is, perhaps, statistics which forms a large part of geo-mathematics and which is well covered by many excellent textbooks. The statistics chapter in this book should form a good introduction to the material covered in those more specialized texts.

Mathematics is much more akin to a language than a science. It is a method of communication rather than a body of knowledge. Thus, the best way to approach a book like this is as you would a text on, say, French or German. You are learning how to communicate with people who understand the mathematical language. You are not learning a collection of facts. Another similarity to learning a language is that you must never pass on to the next lesson until you have grasped the current one. If you do, you will get hopelessly lost and demoralized since succeeding chapters will simply make no sense.

So that you know you have understood sufficiently to move on, each chapter is sprinkled with examples for you to attempt. Mostly these are very short and simple. A few, however, are more difficult and are designed to make you think carefully about the maths just discussed. If you are unable to do one, you should read over the preceding paragraphs again and make sure that you have understood everything. If that does not help then get assistance. Each chapter concludes with additional simple questions as well as more wide-ranging questions which will test your ability to apply what you have learnt to more realistic problems. Outline answers to most questions are given at the end of the book and more complete answers are given for some of the more difficult problems. Look carefully at these complete answers since they also show how your answers should be set out. I assume, throughout this book, that you have a calculator and know how to use it.

One difficulty that many students have with mathematics is the large number of specialized mathematical words. Sometimes these words are completely new to the student whilst other times they are used in a similar, but somewhat more precise, manner to their everyday meaning. It is impossible to avoid use of such words since they are vital in mathematics. Wherever I introduce such a word it is in bold face (e.g. **jargon**).

This first chapter is about basic tools that are needed in succeeding chapters and will introduce you to the most important ideas needed for application of mathematical principles to geological problems.

1.2 MATHEMATICS AS AN APPROXIMATION TO REALITY

Geology is frequently regarded as a **qualitative** (i.e. descriptive) science. Geological discussions often revolve around questions about what happened and in what order. For example, was a particular area under the sea when a given sedimentary rock was deposited and does the erosive surface at the top imply that uplift above sea level occurred subsequently? However, the same geological information can be described **quantitatively** (i.e. by numbers). In the preceding example, how deep was the sea and how long was it before uplift occurred? Geology is also concerned with the influence of one process upon another. How does changing water depth affect sediment type? Once again it is possible to do this quantitatively by producing equations relating, say, grain size to water depth (unlikely to be very accurate but in principle possible).

Figure 1.1 illustrates a situation in which a quantitative description can be attempted. The figure shows a lake within which sediment, suspended in the water, rains down and slowly builds up on the lake floor. Obviously early deposits will be covered by later ones. This results in a relationship between depth below lake bed and time since deposition; the deeper you go the older the sediments get. Now, if the rate at which sediments settle upon the floor is approximately constant, sediments buried 2 m below the lake bed are twice

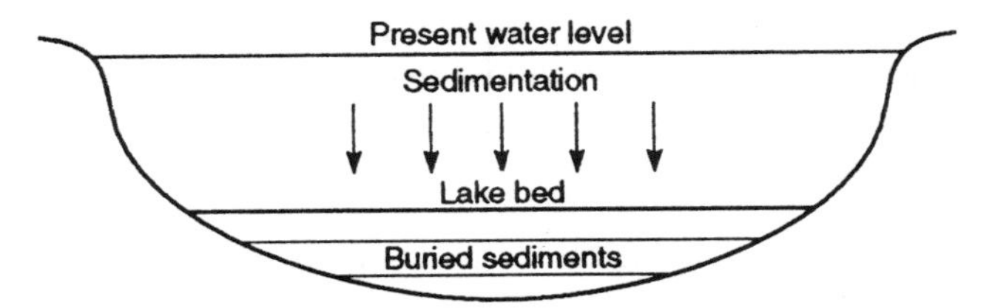

Figure 1.1 Sedimentation on a lake bottom. As sediment accumulates on the lake bed, older sediments are slowly buried by younger deposits.

as old as sediments buried by 1 m and sediments buried by 3 m are three times as old and so on. Thus, if you double depth, you double the age, if you triple depth you triple the age and so on. This means that the sediment age is **proportional** to burial depth. This can be expressed, mathematically, by the equation

$$\text{Age} = k \times \text{Depth} \tag{1.1}$$

where k is a **constant**. In other words, the age of the sediment equals its depth multiplied by a constant. Constants are values which do not change within a given problem. In the case of the problem above, the constant tells us how rapidly sediments accumulate. A large value for k implies that age increases very rapidly as depth increases (i.e. sediments accumulated very slowly). A low value implies that the age increases more slowly (i.e. sediments accumulated more rapidly). In a particular lake it might take 1500 years for each metre of sediment to accumulate. In this case $k = 1500$ years per metre (y m^{-1}). A lake with a lower sedimentation rate of, say, $3000\ \text{y m}^{-1}$, would have a more rapid increase in age with depth of burial.

Question 1.1 If $k = 1500\ \text{y m}^{-1}$ calculate, using equation (1.1), the age of sediments at depths of 1 m, 2 m and 5.3 m. Repeat the calculations for $k = 3000\ \text{y m}^{-1}$.

As you see, it is possible to produce mathematical expressions relating geological **variables** to each other (a variable is a quantity which, in a particular problem, can change its value, e.g. the variable called 'Age' in equation (1.1) changes when the variable called 'Depth' is altered). Are such quantitative descriptions worth bothering with and are the results worth having? Well, it depends! Sometimes such an exercise will not tell you anything you did not already know. On other occasions the ability to manipulate and combine mathematical expressions can lead to new insight into geological processes. Mathematical expressions also have the great merit of consistency; they always give the same answer when you use the same data (unlike some geologists I know). Finally, mathematical expressions are capable of being definitively tested. An expression can be used to predict a result and that result can then be checked. Equation (1.1) could be used to predict the age at a particular depth and that age could be tested using, say, a geochemical dating method. If the age is very wrong then there is something wrong with the geological or mathematical model, e.g. some important factor is missing.

Unfortunately, it is quite possible for mathematics to give the wrong answer. In fact mathematical results are rarely 100% correct. Hopefully though, a mathematical relationship is at least approximately true. The lake sediment example, equation (1.1), is a good case. The assumption that sedimentation happened at the same rate throughout deposition and the, hidden, assumption that the sediments are not compacted by the weight of overlying deposits are unlikely to be completely true. However, provided the sedimentation rate does not vary too much and provided sediment compaction is not too extreme, equation (1.1) should be approximately correct. This is all that is necessary for a mathematical formulation to be useful. It is worth keeping the fact that mathematical expressions are usually approximations at the back of your mind. People often make the assumption that, because a mathematical expression is being used, the answer must be right. This is simply not true, not even in physics (equations in physics are also approximations to reality although the approximation is usually so good that this fact can safely be overlooked).

A final general point about using mathematics for solving problems. If you look on any page in this book, or at any mathematical paper in a geological journal, you will see that there is far more text than there is mathematics. A frequent failing in students' use of mathematics is to write down lots of equations with no explanation of what they are or what they mean. The result is an obscure piece of work which nobody else, even the students themselves six months later, can understand. It is also a recipe for sloppy or illogical mathematics. A good guide is that there should be rather more English in a piece of mathematics than equations. At minimum, you must describe all of the constants and variables that you use. It is also good practice to number all equations since this makes it easier for yourself or anybody else to refer to a particular expression in later discussion. A clear diagram can also be an important part of a good mathematical explanation.

1.3 USING SYMBOLS TO REPRESENT QUANTITIES

The lake sediment equation used the world 'Depth' to represent the quantity depth. However, any other symbol would do equally well. Equation (1.1)

$$\text{Age} = k \times \text{Depth} \tag{1.1}$$

could be written

$$a = kz \tag{1.2}$$

Table 1.1 Lower case and upper case letters of the Greek alphabet. Greek letters are frequently used in mathematical expressions even though use of the equivalent roman letters would be equally valid

Greek characters	*Name*
α, A	alpha
β, B	beta
γ, Γ	gamma
δ, Δ	delta
ε, E	epsilon
ζ, Z	zeta
η, H	eta
θ, Θ	theta
ι, I	iota
κ, K	kappa
λ, Λ	lambda
μ, M	mu
ν, N	nu
ξ, Ξ	xi
o, O	omicron
π, Π	pi
ρ, P	rho
σ, Σ	sigma
τ, T	tau
υ, Y	upsilon
ϕ, Φ	phi
χ, X	chi
ψ, Ψ	psi
ω, Ω	omega

where a is age, k is the sedimentation constant and z is depth. Alternatively, it could be written

$$\alpha = \kappa\zeta \tag{1.3}$$

where α is age, κ is the sedimentation constant and ζ is depth (see Table 1.1 for a list of Greek letters such as those used here). The point is that it really does not matter. Equation (1.3) is just as valid and just as simple as equation (1.1). However, the unfamiliarity of Greek letters can make an equation look rather daunting. The same equation could equally well be written using Hebrew characters, Chinese pictograms, Egyptian hieroglyphics or even some completely new set of symbols. The use of Greek letters extensively throughout mathematics is really just a tradition although it does have the benefit of doubling the number of available symbols.

Also traditional is the use of particular symbols to represent commonly encountered quantities. A good example is z which is nearly always used to

Table 1.2 Commonly used symbols and their usual meanings

Symbol	*Usual meaning*
z	Depth
T	Temperature
t	Time
x	Horizontal distance
ρ	Density
ϕ	Porosity and grain size
θ	An angle
P	Pressure
r	Radius
v	Velocity
σ	Stress

represent depth. A few other common examples are given in Table 1.2 which is far from complete but should give the general idea. These symbols are so commonly used for these particular variables that people frequently forget to define them. Thus if, in particular book or paper, the author is discussing crustal temperatures and the symbol T appears, you can be fairly sure that this will represent a temperature even if the author forgets to define it as such. However, this is not good practice and all symbols should normally be defined.

A few symbols also have specialized mathematical meanings such as the Greek letter delta (i.e. Δ or δ) which is used to denote a small change in a variable. If temperature in the lower crust increased, due to some thermal event, by a small amount (say 10 °C) this temperature change is given the symbol ΔT (or δT). If the original temperature was T °C the increased temperature is then $(T + \Delta T)$ °C. Another, well-known, example of reserving a symbol for a particular mathematical purpose is the use of π to denote the number 3.141 59...; again this is such common usage that you will virtually never see π defined in a book or a paper. In this case, however, this is acceptable since this convention is universally adopted throughout mathematics and the sciences. Other examples of specialized mathematical meanings for symbols will be covered as we come across them in this book.

1.4 SUBSCRIPTS AND SUPERSCRIPTS

Another feature of mathematical expressions, which some people find confusing, is the use of subscripts and superscripts. Subscripts are usually used to qualify the meaning of a symbol. For example, if T is used in an expression to denote temperature as a function of depth in the earth, then T_0 might well be used to denote the temperature at the earth's surface (i.e. at depth = 0.0 m).

Similarly, if a sandstone and a shale are being compared, their densities might be given the symbols ρ_{sand} and ρ_{shale} respectively and their porosities would be given the symbols ϕ_{sand} and ϕ_{shale}. The use of subscripts is no different from the use of any other strange character to denote a quantity. Subscripts simply clarify the meaning of a particular symbol.

Superscripts (often called **exponents**) have a definite mathematical meaning. A superscript is an instruction to raise a number to a power. Thus a^2 means 'square a', a^3 means 'find the cube of a' and a^n means 'multiply a by itself n times'. Whilst on the subject of raising numbers to a power, it is worth briefly reviewing a couple of simple points that will be needed later on. There are three manipulations in particular that you should be familiar with:

$$x^a x^b = x^{a+b} \tag{1.4}$$

$$x^a / x^b = x^{a-b} \tag{1.5}$$

and

$$(x^a)^b = x^{ab}. \tag{1.6}$$

For example, 100 ($= 10^2$) times 100 equals 10 000 ($= 10^4$), i.e.

$$10^2 \times 10^2 = 10^{2+2} = 10^4$$

and the cube of 4 ($4 = 2^2$) is 64 ($64 = 2^6$), i.e.

$$(2^2)^3 = 2^{2\times 3} = 2^6.$$

Question 1.2 Simplify and, where possible, evaluate the following expressions:

(i) $5^2 \times 5^4$, (ii) $(5^2)^4$, (iii) $x^2 \times x^3$, (iv) $\text{Depth}^2 \times \text{Depth}^3$, (v) $(T_0^3)^4$ where $T_0 = 10$.

1.5 VERY LARGE NUMBERS AND VERY SMALL NUMBERS

Many quantities in geology are very large (e.g. the mass of the earth) or very small (e.g. the mass of gold suspended in one litre of sea water). It is therefore vital that you understand how to deal with very small or very large numbers. Two ways of talking about such extreme numbers are: (i) the use of **scientific notation** (or standard form), and (ii) the use of special, very large or very small, units. Methods are also required for specifying very small fractions such as the fraction of a rare element contained in a mineral sample.

Table 1.3 Large numbers of various sizes expressed as a power of 10

Number	*Power of 10*
1 000	10^3
10 000	10^4
100 000	10^5
1 000 000	10^6
1 billion	10^9

Scientific notation is the most flexible of the methods for discussing the very large and the very small. Table 1.3 shows how various large numbers can be represented by powers of 10. Note that in this book, and in most scientific literature, the American usage for one billion (one thousand million) is used rather than the British norm (one million million). The quick way to find out which power of 10 to use for a particular number is simply to count the number of zeros. One million is 1 followed by six zeros and therefore equals 10^6.

This is fine for giving large numbers which are an exact power of 10. What about numbers such as two million? This is easy; two million is two times one million. This number is therefore written

$$\begin{aligned} 2\,000\,000 &= 2 \times 1\,000\,000 \\ &= 2 \times 10^6 \end{aligned}$$

This is an example of scientific notation for a large number. Other more complex numbers can also be dealt with such as 2 200 000. This is simply 2.2 times one million and can therefore be written as 2.2×10^6. The same number could equally well be thought of as 22 times 100 000 leading to 2 200 000 = 22×10^5. However, for scientific notation it is usual to have the multiplier falling between 1 and 10 and so the former expression (i.e. 2.2×10^6) is preferred.

Question 1.3 Express the following numbers in scientific notation:

(i) 1000, (ii) 2000, (iii) 2500, (iv) 2523, (v) 23 000 000, (vi) seven billion.

Table 1.4 Small numbers of various sizes expressed as a power of 10

Number	*Power of 10*
0.001	10^{-3}
0.0001	10^{-4}
0.000 01	10^{-5}
0.000 001	10^{-6}
1 billionth	10^{-9}

Small numbers can be dealt with in a similar manner. Table 1.4 shows how various small numbers are expressed as powers of 10. Again there is a quick way of determining the power of 10 to use. Count the zeros including the zero before the decimal point. For example, 0.0001 has four zeros in total and is written as 10^{-4}. Do not worry if it is not clear to you why a negative power of 10 can be used to express these small numbers, this will be explained further in Chapter 2. For now it is only necessary that you accept that it works. Extending this system for numbers which are not exactly a power of 10 is then achieved by introducing a multiplier. For example, 0.0002 is twice 0.0001 giving

$$\begin{aligned} 0.0002 &= 2 \times 0.0001 \\ &= 2 \times 10^{-4}. \end{aligned}$$

Another example is 0.000 0054 which is written

$$\begin{aligned} 0.000\,0054 &= 5.4 \times 0.000\,001 \\ &= 5.4 \times 10^{-6}. \end{aligned}$$

Question 1.4 Express the following numbers in scientific notation:

(i) 0.001, (ii) 0.002, (iii) 0.0025, (iv) 0.002 523, (v) 0.0 000 023, (vi) seven billionths.

SI (Système International) units are an alternative to the use of scientific notation. In this system a new unit is introduced for each thousandfold increase or decrease in size. For example, the basic unit of distance is the metre (m). The next unit up from this is the kilometre (km) which is 1000 times bigger. One thousand kilometres is one million metres and this unit is denoted the megametre (Mm). Continuing on up the sequence we get to one billion metres (the gigametre, denoted Gm) and to one million million metres (the terametre, denoted Tm). Moving in the opposite direction, one thousandth of a metre is a millimetre (mm) and one millionth of a metre is a micrometre (μm). Finally, we get to one thousand millionth of a metre which is called a nanometre (nm). Exactly the same set of prefixes is used for any other SI unit. Thus the mass units, starting from the very small and increasing one thousandfold for each step, are the nanogram (ng), the microgram (μg), the milligram (mg), the gram (g), the kilogram (kg), the megagram (Mg), the gigagram (Gg) and the teragram (Tg). These prefixes, and a few extra ones, are summarized in Table 1.5.

It is worth noting that a frequently encountered error in the use of this system is to use 'K' rather than 'k' in, for example, kilometre (i.e. this is written km not Km). A capital K is reserved for the **Kelvin** scale of temperature and thus 'K m' is an abbreviation of 'Kelvin metres', not 'kilometres'.

Apart from common usages such as kilometre and kilogram, this method for discussing the very large or the very small is not widely employed in geology. One exception is the use of ky and My for thousands of years and

Table 1.5 List of prefixes used in the SI system for denoting very small and very large units. Each prefix represents a unit 1000 times larger than the prefix on the preceding line. Thus a millimetre is 1000 times larger than a micrometre and a femtogram is 1000 times smaller than a picogram

Multiple	*Prefix*	*Symbol*	*Example*
10^{-18}	atto	a	attometre (am)
10^{-15}	femto	f	femtometre (fm)
10^{-12}	pico	p	picometre (pm)
10^{-9}	nano	n	nanometre (nm)
10^{-6}	micro	μ	micrometre (μm)
10^{-3}	milli	m	millimetre (mm)
1	No prefix		metre (m)
10^{3}	kilo	k	kilometre (km)
10^{6}	mega	M	megametre (Mm)
10^{9}	giga	G	gigametre (Gm)
10^{12}	tera	T	terametre (Tm)

millions of years respectively. (N.B. ka and Ma are also frequently used to denote thousands of years and millions of years.)

Question 1.5 How long, in years, is 31.6 gigaseconds? (Hint: first work out how many seconds there are in a year of 365.26 days.) Using scientific notation, how many seconds is this?

When it comes to denoting very small fractions, the usual approach is a simple extension of the percentage system. In percentage notation, the figure '23%' means 23 parts in every hundred. Thus, if a rock specimen is 23% iron by weight, it contains 23 g of iron in every 100 g of rock. However, this is not a convenient system for discussing trace elements present in very small fractions of a per cent. Small proportions can be represented by talking about parts per million (ppm) or parts per billion (ppb). The rock specimen might contain 17 ppb of the element lanthanum. This means that every billion grams will contain 17 g of lanthanum. It might also contain, say, 10 ppm of gold. Thus every million grams of rock will contain 10 g of gold (i.e. 10 g of gold in every metric tonne of rock).

Question 1.6 Express 0.01% in ppm.

1.6 MANIPULATION OF NUMBERS IN SCIENTIFIC NOTATION

Scientific notation is used frequently both in this book and throughout geological literature. You therefore have to know how to add, subtract, multiply and divide numbers expressed in this way.

The trick with addition or subtraction is to use the same power of 10 for all numbers. For example, the net rate of increase in mountain height is given by the rate of uplift, which increases mountain height, minus the rate of erosion which tends to reduce mountain height (Figure 1.2). If the rate of uplift is $3 \times 10^{-3}\,\mathrm{m\,y^{-1}}$ whilst the rate of erosion is $5 \times 10^{-4}\,\mathrm{m\,y^{-1}}$, the net rate of increase in the mountain height is

$$\begin{aligned}\text{Rate of increase in height} &= \text{Rate of uplift} - \text{Rate of erosion} \\ &= (3 \times 10^{-3}) - (5 \times 10^{-4})\end{aligned} \tag{1.7}$$

The problem here is that the first number has an exponent of -3 whilst the second has an exponent of -4. However, the rate of erosion can be expressed with an exponent of -3 as follows:

$$5 \times 10^{-4} = 0.5 \times 10^{-3}.$$

Note that the 5 has been reduced by a factor of 10 (to give 0.5) whilst the 10^{-4} has been increased by a factor of 10 (to give 10^{-3}). Thus the overall effect is to leave the value unchanged. Replacing the rate of erosion by this new expression gives

$$\text{Rate of increase in height} = (3 \times 10^{-3}) - (0.5 \times 10^{-3}).$$

Once the numbers have been expressed using the same power of 10, the subtraction can be performed:

$$\text{Rate of increase in height} = 2.5 \times 10^{-3}\,\mathrm{m\,y^{-1}} \quad (\text{since } 3 - 0.5 = 2.5).$$

Question 1.7 Evaluate the following:

(i) $(2.5 \times 10^{9}) + (1.5 \times 10^{9})$, (ii) $(2.5 \times 10^{9}) + (1.5 \times 10^{8})$, (iii) $(2.5 \times 10^{7}) - (1.5 \times 10^{7})$, (iv) $(2.5 \times 10^{14}) - (1.5 \times 10^{13})$.

Multiplication and division are much more straightforward. The trick here is to rearrange the expressions so that all the multipliers are together and all the

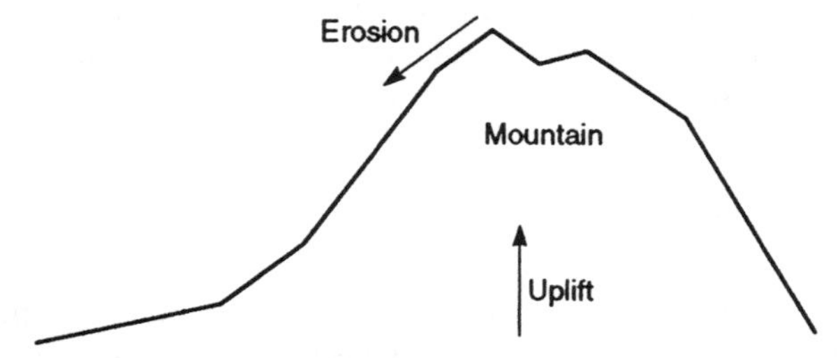

Figure 1.2 The net rate of rise of a mountain equals the rate at which the mountain rises due to uplift minus the rate at which the mountain height drops due to erosion of the peak.

powers of 10 are together. An example should illustrate this:

$$(2.5 \times 10^4) \times (3.0 \times 10^3)/(7.5 \times 10^6) = (2.5 \times 3.0/7.5) \times (10^4 \times 10^3/10^6)$$

where, at this point, no calculation has been performed. The multipliers and powers of 10 have simply been collected together. The calculations implied by the two bracketed terms can then be evaluated to give

$$\begin{aligned}(2.5 \times 3.0/7.5) \times (10^4 \times 10^3/10^6) &= (7.5/7.5) \times (10^7/10^6)\\ &= 1.0 \times 10^1\\ &= 10.0\end{aligned}$$

Question 1.8 If the mass of the earth is 5.95×10^{24} kg and the volume is 1.08×10^{21} m^3, calculate the average density. (Note that density is mass divided by volume.)

I will finish this section with a few words of warning about using calculators for performing these sorts of calculations. There are two problems which this frequently causes. Firstly, many students will write down the results in a similar way to the manner in which they appear on the calculator display. Thus, if the correct answer is 3.01×10^8, this appears in the calculator something like [3.01 ⁸]. There is a strong temptation to write this down as 3.01^8 which means 3.01 to the power 8 rather than 3.01 times 10 to the power 8. The second problem is that many students would enter this number by typing the following buttons:

[3], [.], [0], [1], [×], [1], [0], [exp], [8]

which gives a display reading [3.01^9]. This is because the correct sequence of buttons should have been

[3], [.], [0], [1], [exp], [8]

since the [exp] button means multiply by 10 to the power of the following number. Try entering 3.01×10^8 in the two ways suggested above; you should see what I mean. In general, I would strongly recommend that you perform such calculations as shown in the earlier examples, i.e. separate the multiplier and exponent terms and then use a calculator only for the multiplier part (you should be able to do the exponent part in your head).

1.7 USE CONSISTENT UNITS

Whenever a calculation is performed, all values used must be expressed using the same units. For example, in the calculation given above for finding the rate of rise of a mountain

$$\text{Rate of increase in height} = \text{Rate of uplift} - \text{Rate of erosion} \tag{1.8}$$

the rate of uplift and the rate of erosion must be given using the same units. Thus if the rate of uplift was given as

$$\text{Rate of uplift} = 3 \times 10^{-3}\,\text{m}\,\text{y}^{-1}$$

whilst the rate of erosion was given as

$$\text{Rate of erosion} = 1\,\text{m}\,\text{ky}^{-1}$$

the calculation cannot be performed using these figures since the first figure has units of metres per year whilst the second has units of metres per thousand years. One of the two figures must be converted to the form of the other. In this case it is probably easiest to rewrite the rate of uplift as

$$\text{Rate of uplift} = 3\,\text{m}\,\text{ky}^{-1}$$

which is the same as $3 \times 10^{-3}\,\text{m}\,\text{y}^{-1}$ since the amount of uplift in 1000 years is 1000 times more than that in one year (i.e. $1000 \times (3 \times 10^{-3}) = 3$). Equation (1.8) can then be evaluated to give a net rate of mountain rise equal to $2\,\text{m}\,\text{ky}^{-1}$.

Similarly, in the lake sedimentation problem (equation (1.1)) all figures must have consistent units. Thus, if the age is quoted in years and the depth is quoted in metres, the sedimentation constant will have units of years/metre. If a depth is given in centimetres whilst k is given in years/metre, the depth must first be converted to metres before the calculation of age is done.

Question 1.9 Using equation (1.1) and a sedimentation constant of $1000\,\text{y}\,\text{m}^{-1}$, find the age of sediment buried at a depth of 30 cm.

FURTHER QUESTIONS

1.10 If $\Omega_d = 3.1 \times 10^4$ and $\mu = 2.7 \times 10^{-2}$ evaluate $\Delta J = \mu/\Omega_d$.

1.11 The earth gains mass every day due to collision with (mostly very small) meteors. Estimate the increase in the earth's mass since formation assuming that the rate of collision has been constant and that

$$\Delta M = 6 \times 10^5\,\text{kg}\,\text{day}^{-1}$$

$$A_e = 4.5 \times 10^9\ \text{years}$$

where ΔM is the rate of mass gain and A_e is the age of the earth. What is this mass gain as a fraction of the total present mass of the earth, M_e, where

$$M_e = 5.95 \times 10^{24}\,\text{kg}.$$

Re-express this answer in ppb.

1.12 Calculate the volume of the earth using the expression

$$V = \frac{4\pi r^3}{3}$$

where r is the earth's radius (equal to 6.37×10^6 m). Note that this method assumes that the earth is a perfect sphere.

1.13 How long would it take to travel 100 km at 20 km hour^{-1}? The following problem is identical in form: the North Atlantic Ocean is getting wider at an average rate, v_s, of around 4×10^{-2} m y^{-1} and has a width, w, of approximately 5×10^6 m.

(i) Write an expression giving the age, A, of the North Atlantic in terms of v_s and w assuming the present-day spreading rate is typical of the ocean's entire history.
(ii) Evaluate your expression by substituting the values given above.

Common relationships between geological variables

2

2.1 INTRODUCTION

This chapter is about relationships between variables. In the last chapter, the depth and age of sediments in a lake were related by the simple expression

$$\text{Age} = k \times \text{Depth}. \qquad (1.1)$$

Many other geological variables are also related to each other. For example, the internal temperature of the earth is related to depth (it gets hotter as you get deeper) and the strength of a rock is related to the pressure applied to it (rocks usually become stronger when compressed). However, the precise nature of the relationship varies from one example to another. For simple cases expressions similar to equation (1.1) will do. In other cases the relationships are more complex.

In equation (1.1) above, age is a **function** of depth. This implies that any given depth produces a unique value for the age. Any type of relationship in which the value of one variable produces a single, unique, value for another is called a function and this term will be met with repeatedly throughout the remainder of this book.

This chapter is probably the most important in the entire book. It is only the fact that we can use mathematical expressions to relate different geological quantities that makes mathematics useful in geology. In practice, the true relationships between quantities such as depth and temperature are usually so complex that they must be approximated by much simpler ones. This chapter will introduce you to some of the most common of these simple relationships starting with the most simple and common of all, the straight line function.

2.2 THE STRAIGHT LINE

The straight line equation is possibly the most important mathematical expression found in geology since a very large range of geological problems can be approximated using straight line functions.

Returning to the lake sediment problem of Chapter 1, imagine that the lake completely dried out 1 My ago and that there has been no sedimentation in the lake since that time. Under these circumstances all the sediments are 1 My older than we might otherwise think, i.e. the top sediments are 1 My old rather than recent and sediments at a depth of 1 m are, say, 1 000 500 years old rather than 500 years old. An equation to describe the age of the sediments would now be

$$\text{Age} = (k \times \text{Depth}) + \text{Age of top} \qquad (2.1)$$

since, in this expression, each age calculated from '$k \times$ Depth' has the 'Age of top' added to it. Thus, if the sedimentation constant k was 500 y/m and the age of the top sediments is 1 My the sediments have an age of

$$\text{Age} = (500 \times \text{Depth}) + 1\,000\,000. \qquad (2.2)$$

Sediments buried at a depth of 100 m would then have an age of

$$\begin{aligned} \text{Age} &= (500 \times 100) + 1\,000\,000 \\ &= 1\,050\,000 \\ &= 1.05\ \text{My}. \end{aligned}$$

Repeating this calculation for a range of depths from 0 to 100 m gives the values shown in Table 2.1 and plotted in Figure 2.1.

Question 2.1 Repeat the above calculation for a depth of 50 m

As you can see, the resultant graph is a straight line. A straight line is completely specified by just two quantities. Firstly, the point where the line

Table 2.1 The ages of buried sediments in a dried-out lake bed if sediments accumulated at a rate of 500 years for each metre of sediment and if the lake dried out 1 My ago

Depth (m)	*Age (My)*
0	1.00
20	1.01
40	1.02
60	1.03
80	1.04
100	1.05

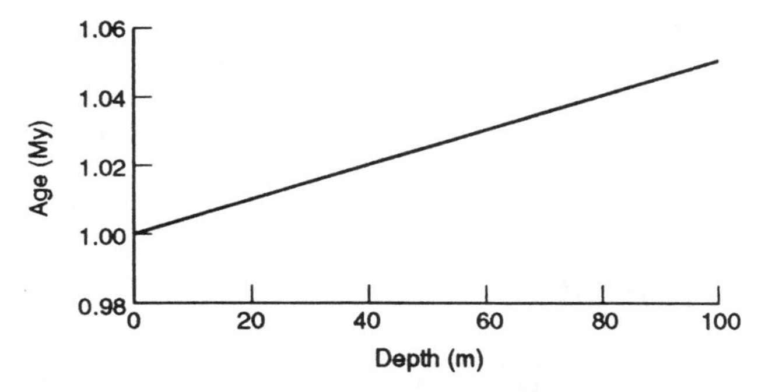

Figure 2.1 Graph of age versus depth data from Table 2.1.

crosses the vertical axis tells us 'how high up' the line is. Secondly, the steepness of the line. A different straight line will either cross the vertical axis at another place or it will be less (or more) steep.

The position where the line crosses the vertical axis is called the **intercept** and has a value of 1 My in the particular case of Figure 2.1. It is essential, when determining this intercept, that the vertical axis crosses the horizontal axis at the point where the depth is zero. If the vertical axis is anywhere else the age at which the plotted line crosses the axis will be different (Figure 2.2). Thus, to specify 'how high up' the straight line is, it would be necessary to give both the intercept and the location of the vertical axis. To avoid this the intercept is always quoted for a vertical axis which passes through the origin of the horizontal axis.

The second value which characterizes a straight line, the steepness, is called the **gradient** of the line. This is simply a measure of how rapidly the 'height'

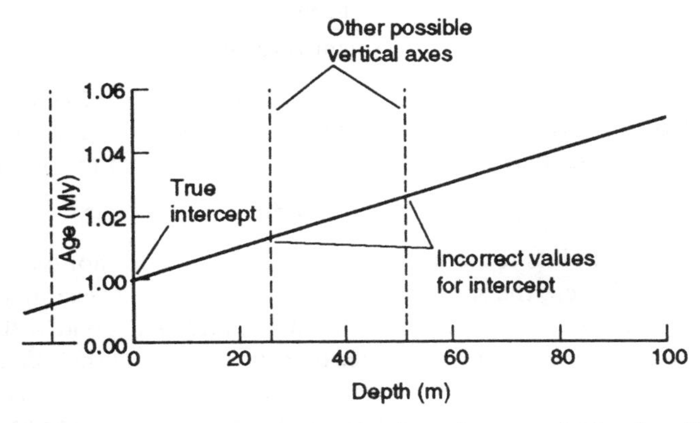

Figure 2.2 The intercept should always be given for a vertical axis which passes through the origin of the horizontal axis. Any other vertical axis will give a different value for the intercept.

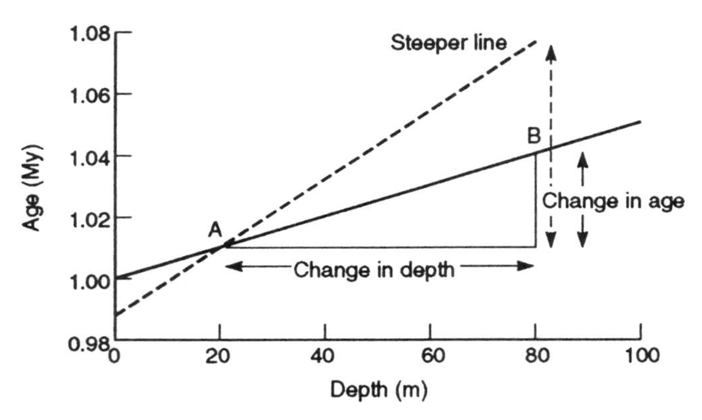

Figure 2.3 Both the depth and age of the sediments alter as we move from point A on the line to point B but note that the steeper the line the more the age changes for a given change in depth.

increases as we move from left to right along the line. The effect of gradient is illustrated by Figure 2.3. Note that both age and depth alter as we go from point A to point B and that, for a given change in depth, the change in age becomes greater as the steepness of the line increases. Thus, the steepness can be characterized by the increase in age produced by a given increase in depth. For simplicity, we can fix the depth increase as being 1 m, i.e. the gradient is defined as the increase in age produced by drilling into the sediments by an additional 1 m. However, the points A and B in Figure 2.3 are not necessarily 1 m apart so we must modify the observed change in age between points A and B by dividing it by the distance between them. Thus, the gradient is given by

$$\text{Gradient} = \frac{\text{Change in age}}{\text{Change in depth}}. \tag{2.3}$$

For example, point A is at a depth of 20 m and an age of 1.01 My whilst point B is at a depth of 80 m and an age of 1.04 My. Thus the change in depth is 60 m and the change in age is 0.03 My (= 30 000 years) giving a gradient of

$$\text{Gradient} = 30\,000/60 = 500\,\text{y}\,\text{m}^{-1}.$$

There are several points to note about this answer. It does not matter which two points are chosen, the same answer would result if, for example, depths of 0 and 50 m had been chosen for the points A and B. Secondly, the 'units' for the answer of 'years/metre' occur because the top line of the calculation is in years (30 000 years) and the bottom line is in metres (60 m). Hence the calculation involves years divided by metres giving an answer in years/metre. This procedure for finding the units of an answer will be covered in more detail in Chapter 3.

Question 2.2 Calculate the gradient of the straight line in Figure 2.3 using the point A again (depth = 20 m, age = 1.01 My) and the point at a depth of 100 m and age of 1.05 My.

The answer of 500 y m^{-1} is not only a measure of the steepness of the line. This gradient tells us that each metre of sediment takes 500 years to accumulate (i.e. 500 years per metre). This is, of course, our sedimentation constant.

It should now be clear that, for the lake sedimentation example, the intercept on a graph of age against depth tells us the age of the top sediments whilst the gradient tells us the rate of accumulation. In fact, rather than obtaining the straight line graph from given values for intercept and gradient, it is more likely that these quantities will be estimated by fitting a straight line to a graph of some depth/age data. Consider the age versus depth data shown in Table 2.2. These figures might, for example, have been obtained by taking cores from a lake bottom and dating them using the radio-carbon method (this is a geochemical method for estimating the age of organic remains, the details of which are beyond the scope of this book). Figure 2.4 shows a graph of these data together with a 'best-fit' straight line which passes very close to all of the points.

Table 2.2 Measured ages of sediments buried at various depths in a lake bed

Depth (m)	*Age (years)*
0.5	1 020
1.3	2 376
2.47	5 008
4.9	10 203
8.2	15 986

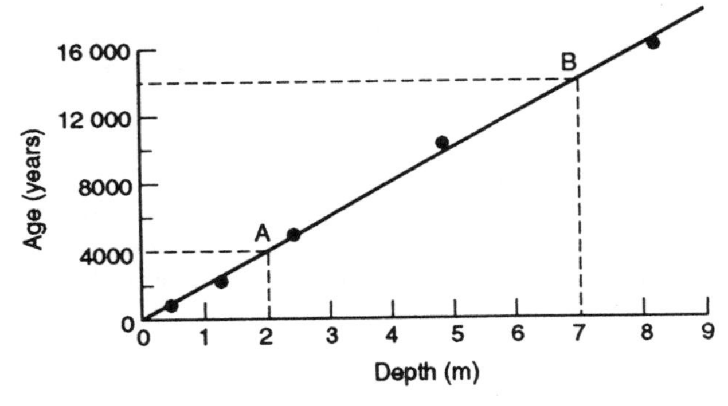

Figure 2.4 A graph of the sediment age versus depth data from Table 2.2.

In this example, the intercept value is not significantly different from zero. Thus, in this lake, sedimentation is continuing at the present day and at the same rate as in the past. The gradient of the line can be found by assessing the points A and B shown. The point A lies at a depth of 2 m and an age of 4000 years whilst the point B has a depth of 7 m and an age of 14 000 years. Thus, the change in depth is 5 m and the change in age is 10 000 years giving a gradient of

$$\text{Gradient} = 10\,000/5 = 2000\ \text{y m}^{-1}$$

i.e. each metre of sediment took 2000 years to accumulate or, equivalently, the sedimentation constant $k = 2000\ \text{y m}^{-1}$.

Question 2.3 Given the following depth/age data from a dried-up lake bed, estimate the rate of sedimentation and how long ago the lake dried out.

Depth (m)	*Age (years)*
6	570 000
10	580 000
18	615 000
20	620 000

It is now time to move from the specific example of lake bottom sedimentation to more general expressions. If we start with our lake sediment equation

$$\text{Age} = k \times \text{Depth} + \text{Age of top} \qquad (2.1)$$

this is not a general equation since the **terms** in equation (2.1) (i.e. 'Age', '$k \times$ Depth', 'Age of top') refer to specific quantities involved in the sedimentation problem. The simplest way to arrive at a more general expression is to replace each of the terms in the equation by new symbols which do not have specific meanings. Thus we can replace the specific variable 'Age' by the general variable y. Similarly, the 'Depth' can be replaced by the general variable x and the constants k and 'Age of top' can be replaced by the general constants m and c. This procedure results in a **general form** for the equation of a straight line of

$$y = mx + c \qquad (2.4)$$

where y is plotted along the vertical axis and x is plotted along the horizontal axis. Note that using the new symbols y, m, x and c was an entirely arbitrary choice. Any other set of symbols could have been chosen. For example, the equation

$$\alpha = \beta\gamma + \delta$$

is also a straight line equation, provided α and γ are variables and β and δ are

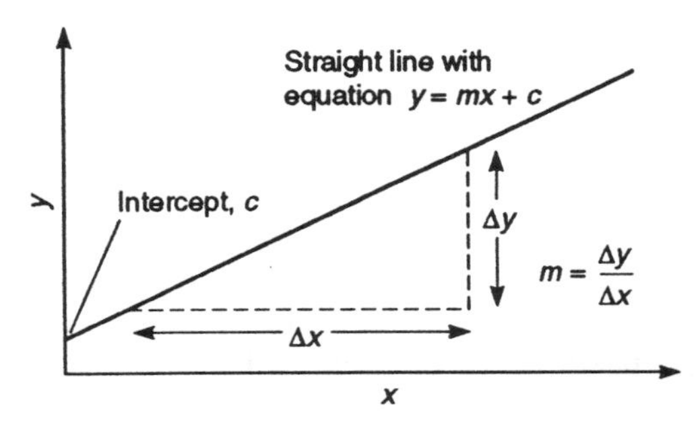

Figure 2.5 The general form of the equation of a straight line, $y = mx + c$.

constants, since it is of the same form as equations (2.1) and (2.4). However, the particular symbols used in equation (2.4) are traditionally used for the general form of the equation of a straight line and I have stuck to that convention.

A graph of equation (2.4) is a straight line which has an intercept of c and a gradient of m (Figure 2.5). In this figure, Δy (pronounced 'delta y', this means a little bit of y) is the change in height produced by moving a horizontal distance Δx along the line. Note that, in the illustrated example, y increases for an increase in x since the graph slopes up to the right. Thus Δy and Δx are positive and so is the gradient. On the other hand, if the line sloped downwards to the right, y would decrease as x increases. Thus, Δy would be negative giving a negative gradient. As a general rule, lines that increase in height towards the right have a positive gradient whilst lines that decrease in height have negative gradients.

One more example may help to make these ideas clearer. It is well known that temperature increases with depth in the earth and, for depths of less than around 100 km, it is a good approximation to assume that a plot of temperature against depth should be a straight line. The intercept of such a graph is, by definition, the temperature at zero depth, i.e. the surface temperature. This value will vary considerably from tropical to polar locations but a typical value might be 10 °C. The gradient of the line, i.e. the rate at which temperature increases with depth, also varies from one location to another since geologically active areas have very different gradients from old, stable, continental areas. However, values around 20 °C km^{-1} are not unusual, i.e. the temperature increases by 20 °C for an increase in depth of 1 km. To summarize, temperature plotted against depth gives a straight line characterized by the local temperature gradient and an intercept equal to the local surface temperature. A general expression for how temperature varies with depth at a particular location would therefore be

$$\text{Temperature} = (\text{Gradient} \times \text{Depth}) + \text{Surface temperature} \qquad (2.5)$$

(cf. $y = mx + c$). For the specific case of an intercept of 10 °C and a gradient of 20 °C km^{-1} this would yield

$$\text{Temperature} = (20 \times \text{Depth}) + 10.$$

Thus, at a depth of 40 km the temperature is 810 °C.

> Question 2.4 Rocks usually increase in strength, τ, when compressed. This strength is defined as the shearing (= sideways) pressure necessary for a particular rock specimen to break. The standard units of pressure are pascals. If τ increases by m pascals for each additional pascal of normal pressure (i.e. compressive pressure) and if the strength when not compressed is τ_0, write an equation for how τ varies with normal pressure P. Sketch a graph of this function.

There are many other examples in geology of the use of straight line functions. However, many geological phenomena are not well represented by straight lines and more complex expressions must be used. Some of the more common alternatives are described in the remainder of this chapter.

2.3 QUADRATIC EQUATIONS

The linear temperature with depth relationship discussed in the last section breaks down badly for depths much larger than 100 km. For example, at the centre of the earth the depth is approximately 6360 km so that a surface temperature of 10 °C and a gradient of 20 °C km^{-1} would predict a temperature of

$$\begin{aligned} \text{Temperature} &= (\text{Gradient} \times \text{Depth}) + \text{Surface temperature} \qquad (2.5)\\ &= (20 \times 6360) + 10 \\ &= 127\,210\,^{\circ}\text{C}. \end{aligned}$$

In fact, the temperature in the earth's core is only about 4300 °C! Table 2.3

Table 2.3 The temperature at various depths in the earth as determined from geophysical and geochemical measurements

Depth (km)	*Temperature (°C)*
0	10
100	1150
400	1500
700	1900
2800	3700
5100	4300
6360	4300

lists the approximate temperature at various depths, based upon geophysical and geochemical measurements.

The problem is that the temperature near the surface rises much more rapidly than it does deeper in the earth, e.g. over 1000 degrees in the first 100 km but only 350 degrees in the following 300 km. Note that the temperature is virtually constant within the inner core (i.e. from 5100 km to the earth's centre). Any attempt to extrapolate down to the core using the large rate of increase in temperature near the surface is bound to give a ridiculously large value. A much better approximation is

$$\text{Temperature} = (-8.255 \times 10^{-5})z^2 + 1.05z + 1110 \tag{2.6}$$

where z is the depth in kilometres. This equation contains three terms (i.e. $(-8.255 \times 10^{-5})z^2$, $1.05z$ and 1110) each of which is calculated separately before adding them together. Thus at a depth $z = 5100$ km the temperature is

$$\begin{aligned}\text{Temperature} &= (-8.255 \times 10^{-5} \times 5100 \times 5100) + (1.05 \times 5100) + 1110 \\ &= -2147 + 5355 + 1110 = 4318\,^\circ\text{C}\end{aligned}$$

which compares well with the value given in Table 2.3. However, at the earth's surface, equation (2.6) predicts

$$\begin{aligned}\text{Temperature} &= (-8.255 \times 10^{-5} \times 0 \times 0) + (1.05 \times 0) + 1110 \\ &= 1110\,^\circ\text{C}\end{aligned}$$

which is certainly not correct. In fact, equation (2.6) is a reasonable approximation to the earth's internal temperature only for depths greater than around 100 km.

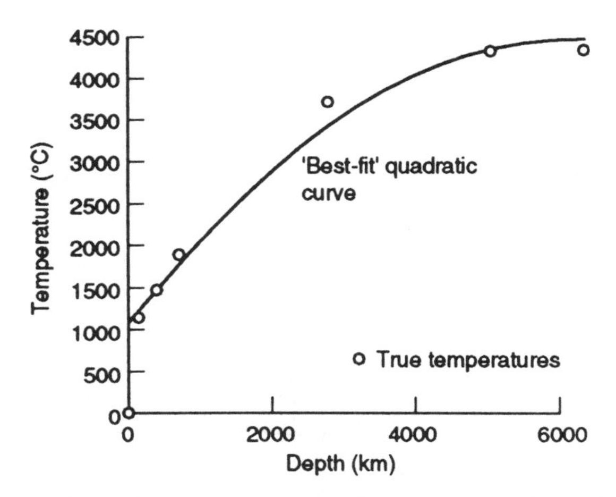

Figure 2.6 Comparison of the earth's internal temperature determined from geophysical and geochemical measurements with the quadratic curve

$$\text{Temperature} = (-8.255 \times 10^{-5})z^2 + 1.05z + 1110$$

where z is depth in kilometres.

Figure 2.6 shows how the temperature variation predicted by equation (2.6) compares with the temperatures given in Table 2.3. Although the fit is not exact, it is clear that equation (2.6) can be used to calculate an approximate value for the temperature at any given depth below 100 km. Once again we see that mathematical descriptions of geological behaviour are useful approximations rather than exact 'truths'.

Equation (2.6) is a particular example of a **quadratic equation**. The general form for this is

$$y = ax^2 + bx + c \tag{2.7}$$

where y is a function of x whilst a, b and c are constants. Figure 2.7 shows a selection of specific examples of equation (2.7).

Comparing the general equation (equation (2.7)) with the temperature profile function (equation (2.6)) we can see that y is equivalent to temperature and x is equivalent to depth. In addition, the constants a, b and c have the values: $a = -8.255 \times 10^{-5}$; $b = 1.05$ and $c = 1110$. The ability to compare a particular equation to a standard form and pick out the appropriate values for the constants will be used again in this book so make sure that you fully understand what has just been done.

> Question 2.5 If
>
> $$f = 2g^2 - 10g + 6$$
>
> where f and g are variables, write down values for the constants equivalent to a, b and c in equation (2.7).

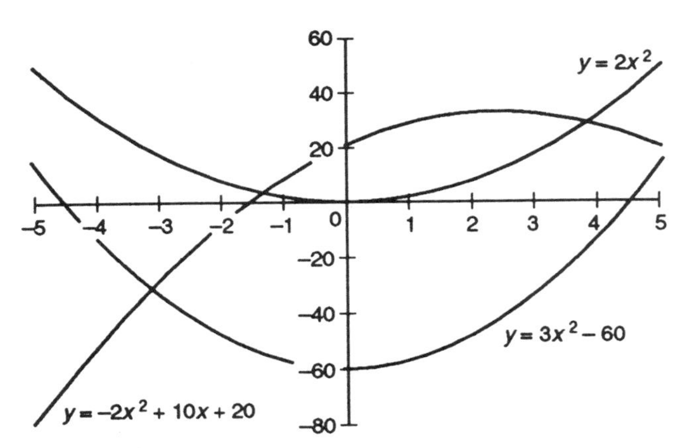

Figure 2.7 Examples of quadratic functions. Each of these curves has an equation of the form $y = ax^2 + bx + c$ but with differing values for the constants a, b and c. Thus, $y = 2x^2$ has $a = 2$, $b = c = 0$ whilst $y = -2x^2 + 10x + 20$ has $a = -2$, $b = 10$ and $c = 20$.

2.4 POLYNOMIAL FUNCTIONS

It is possible to improve further on the fit of a mathematical expression to our temperature data by using longer expressions. Figure 2.8 compares the temperature data to the function

$$\text{Temperature} = az^4 + bz^3 + cz^2 + dz + e \tag{2.8}$$

with values for the constants of $a = -1.12 \times 10^{-12}$, $b = 2.85 \times 10^{-8}$, $c = -0.000\,310$, $d = 1.64$ and $e = 930$. However, whilst the fit to the data is now much better, particularly between 2000 and 4000 km, the expression itself is becoming more difficult to evaluate. This trade-off between accuracy and simplicity is a frequent occurrence when applying mathematics to specific problems. Note that even this more complex expression does not model the temperature in the shallowest 100 km.

> Question 2.6 Compare the temperature predicted by equations (2.6) and (2.8) at a depth of 2800 km. How do these results compare with the true value in Table 2.3?

Expressions like equation (2.8) are known as polynomials (or power series). The general form for these is

$$y = a_0 + a_1x + a_2x^2 + a_3x^3 + \cdots + a_nx^n \tag{2.9}$$

in which '...' indicates a number of terms which have not been explicitly written down. In this expression, a_0, a_1, a_2 etc. are constants and n is an integer

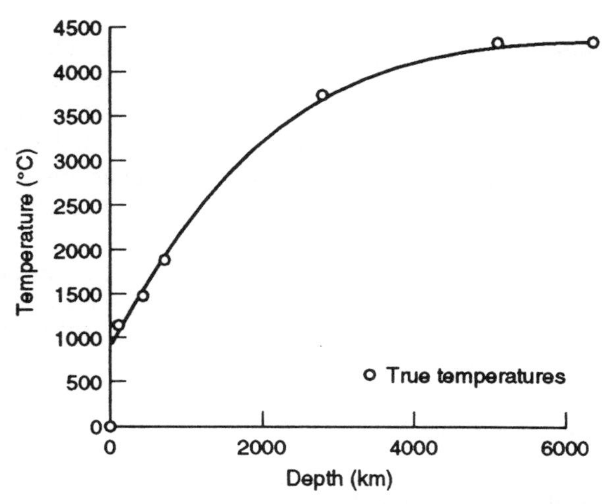

Figure 2.8 A comparison of the temperature data with the function:

$$\text{Temperature} = az^4 + bz^3 + cz^2 + dz + e.$$

giving the power of the last term. For example, if $n = 2$, the expression simplifies to the quadratic function

$$y = a_0 + a_1x + a_2x^2$$

and, if $n = 1$, the expression is a straight line function

$$y = a_0 + a_1x.$$

Thus straight line functions and quadratic functions are special cases of polynomial functions. If $n = 5$, the resulting expression is

$$y = a_0 + a_1x + a_2x^2 + a_3x^3 + a_4x^4 + a_5x^5.$$

2.5 NEGATIVE POWERS

The functions discussed above can be made yet more general by the addition of negative powers. The easiest way to understand these is to note that, for example, 3^4 is 3^5 divided by 3, 6^5 is 6^6 divided by 6 and, in general, x^n is x^{n+1} divided by x. In other words, each decrease of the power by one is achieved by division by x. This is a special case of equation (1.5),

$$x^a/x^b = x^{a-b}, \tag{1.5}$$

in which $b = 1$. Thus, equation (1.5) becomes

$$x^a/x = x^{a-1}.$$

Now, $x/x = 1.0$ since any number divided by itself is 1.0. In addition, from the discussion above, $x/x = x^{1-1} = x^0$. Thus, $x^0 = 1.0$. This is a very important result: any number raised to the power of zero equals one. (The only exception is $0^0 = 0$.) For example, $2^0 = 1$, $100^0 = 1$, $(-36.4)^0 = 1$ and even $\pi^0 = 1$.

This process can then be taken a stage further by division of x^0 by x to give x^{-1}. This is the same as dividing 1 by x, i.e. $x^{-1} = 1/x$. Further divisions lead to $x^{-2} = 1/x^2$, $x^{-3} = 1/x^3$, etc. In other words, a number raised to a negative power equals the reciprocal of the same number raised to a positive power. Thus, $(3.5)^{-96} = 1/(3.5)^{96}$, and, in general, $x^{-n} = 1/x^n$.

We are now in a position to see why, in the scientific notation introduced in Chapter 1, numbers smaller than one are expressed using a negative power of 10. Thus, 0.001 is written as 10^{-3} because it equals $1/10^3$ ($= 1/1000$).

2.6 FRACTIONAL POWERS

The last stage in the generalization of the use of powers is to allow the exponent to be a fraction. Table 2.4 and Figure 2.9 should help to make this idea more acceptable. Table 2.4 lists the result of raising 3 to the power of

Table 2.4 The result of raising 3 to the power of the integers between -2 and 2

n	3^n
-2	1/9
-1	1/3
0	1
1	3
2	9

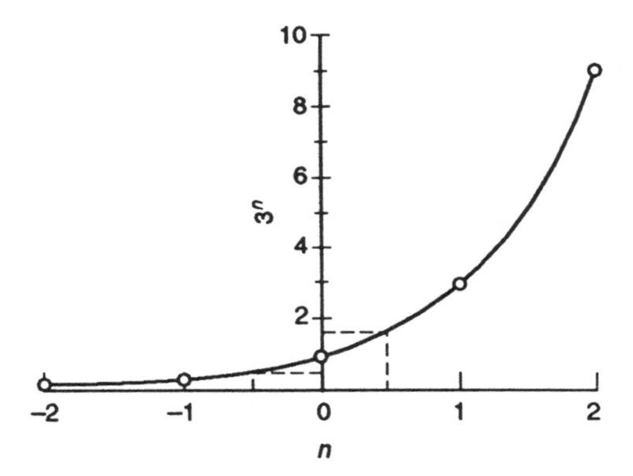

Figure 2.9 A smooth curve drawn through the points in Table 2.4. The dashed lines indicate the point where $n = 0.5$ giving a value for $3^{1/2}$ of approximately 1.7 as well as the point where $n = -0.5$.

the integers between -2 and $+2$. In Figure 2.9 these points are plotted as a simple graph which shows that they lie along a smooth curve. This curve can readily be used at points other than $n = -2, -1, 0, 1$ or 2. For example, at $n = 0.5$ the value on the vertical axis is about 1.7. Similarly, at $n = 1.5$ the vertical axis reads approximately 5.2. Thus $3^{0.5}$ is approximately 1.7 and $3^{1.5}$ is about 5.2. A more mathematically formal treatment of this subject is beyond the scope of this book but the main point to learn here is that it is not necessary to use integers when raising a number to a power. Negative fractional exponents are also possible. From Figure 2.9 it can be seen that $3^{-0.5}$ is around 0.6.

Fractional powers behave in exactly the same way as integer powers. Thus, they obey the equations given in Chapter 1 for manipulating powers, i.e.

$$x^a x^b = x^{a+b} \tag{1.4}$$

$$x^a / x^b = x^{a-b} \tag{1.5}$$

and

$$(x^a)^b = x^{ab}. \tag{1.6}$$

For example, from equation (1.4), $x^{0.3} \times x^{0.4} = x^{0.7}$.

A direct result of this is that some of these fractional powers have a very simple interpretation. For example, a number raised to the power of 0.5 is the square root of the number ($x^{1/2} = \sqrt{x}$) since $x^{1/2} \times x^{1/2} = x^1$. Similarly a number raised to the power of one-third results in the cube root ($x^{1/3} = \sqrt[3]{x}$). Figure 2.9 indicated that $3^{1/2}$ was around 1.7; in fact the square root of 3 is 1.732.

Question 2.7 Draw up a table of 5^n for $n = -2, -1, 0, 1$ and 2. Plot the result. Hence, estimate $1/\sqrt{5}$. In fact, this can be done in two ways. First, estimate it directly from the graph. Secondly, use the graph to estimate $\sqrt{5}$ and then calculate $1/\sqrt{5}$. Compare these answers to each other and to the value given by a calculator.

I will finish this section on polynomial functions and their extensions by using a simple geological example of the use of fractional powers. There are theoretical reasons for expecting water depth, d, in the vicinity of a mid-ocean spreading ridge to depend upon the square root of the distance, x, from the ridge axis according to

$$d = d_0 + ax^{1/2} \tag{2.10}$$

where a is a constant which will depend upon factors such as the spreading rate and d_0 is the depth of the ridge axis. Figure 2.10 shows a comparison between the depths predicted by equation (2.10) and the true water depths in the vicinity of the Pacific–Antarctic spreading ridge assuming a value for d_0 of 2.3 km and a value for a of 0.08. Thus, at a distance from the ridge axis of

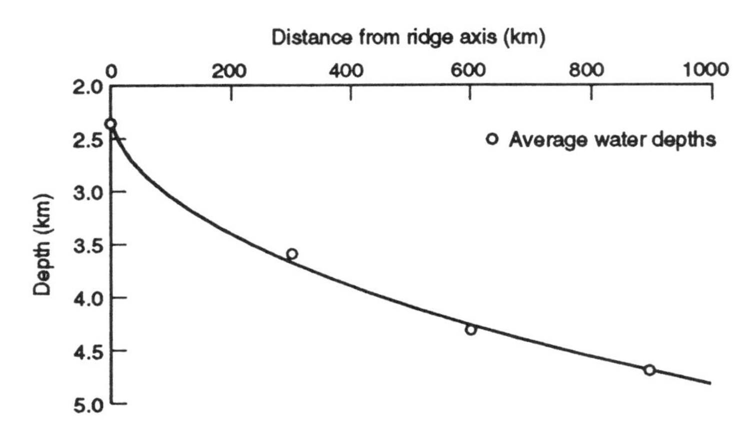

Figure 2.10 Ocean water depth in the vicinity of the Pacific–Antarctic ridge. The solid curve shows how the depth is predicted to vary as a function of distance from the ridge axis.

900 km, equation (2.10) predicts a depth of

$$\begin{aligned} d &= 2.3 + (0.08 \times \sqrt{900}) \\ &= 2.3 + (0.08 \times 30) \\ &= 2.3 + 2.4 \\ &= 4.7\,\text{km}. \end{aligned}$$

As you can see on Figure 2.10, the true water depth is indeed very close to this value.

2.7 ALLOMETRIC GROWTH AND EXPONENTIAL FUNCTIONS

Polynomial functions (section 2.4) and their extensions (sections 2.5 and 2.6) are extremely versatile and can be used to describe many situations. However, there are situations in which they are not appropriate. The way in which sediments compact as they are buried is a good example.

Water contained in recently deposited sediments is usually squeezed out as the sediments are buried. Thus, sediments start with a relatively large porosity and lose this during burial. A particularly simple approximation for the way in which this happens is to assume that a certain proportion of the water is expelled for a given amount of burial. In a particular case half of the water might be released when the sediment is buried by 1 km and half of the remaining liquid removed during further burial to 2 km. If the sediment started with a porosity of 0.6 when deposited, the resulting porosity at various depths would be as shown in Table 2.5.

The important point about this example is that porosity always decreases with increasing burial but never actually reaches zero. It would be very difficult to reproduce this using polynomial functions. However, the values in Table 2.4 could be produced by using

$$\phi = 0.6 \times 2^{-z} \tag{2.11}$$

in which ϕ is the porosity at a depth z (note that porosity is nearly always

Table 2.5 Variation in porosity with depth for a sediment with an initial porosity of 0.6, assuming the porosity decreases by half for each additional kilometre of burial

Depth (km)	*Porosity*
0	0.60
1	0.30
2	0.15
3	0.075
4	0.0375

denoted by the Greek letter ϕ). For $z = 3$ km, 2^{-z} will be $2^{-3} = \frac{1}{8}$ and therefore ϕ becomes $0.6 \times \frac{1}{8} = 0.075$ as shown in Table 2.5.

Question 2.8 What porosity does equation (2.11) give at a depth of 2 km?

Now, whilst equation (2.11) is similar to those discussed in section 2.6, the crucial difference is that the variable, z, appears as the exponent in this expression, i.e. the power used varies as z varies. Compare this to equation (2.10) in which the variable, x, is raised to the fixed power 0.5. This subtle difference produces a rather different type of function. Its general form could be expressed as

$$y = ab^{cx} \tag{2.12}$$

where y is a function of x whilst a, b and c are constants. Equations such as this are called either **allometric growth laws** or **exponential functions**. This equation does not actually need three separate constants since b^c is itself just another constant ($b^c = d$ say) which means that only two independent constants are needed for the general form of this equation. There are two ways of achieving this. First, simply use $b^c = d$ and write equation (2.12) in the form

$$y = ad^x \tag{2.13}$$

which has two constants a and d. Alternatively, the constant b in equation (2.12) is fixed to be a particular, convenient, value and c is retained as an independent constant. For example, $b = 10$ may be simple to use in some contexts leading to expressions like

$$y = a \times 10^{cx} \tag{2.14}$$

The choice of value to use for b is entirely arbitrary and can be varied to suit the problem. However, 99% of the time a rather peculiar choice for b is made. Normally the value $b \approx 2.718$ is used! This number, denoted by the letter e, is special for reasons which will be touched upon in Chapter 8. For now, it is sufficient to know that it is a very important number in mathematics. It is similar to the number π in that it is **irrational** (i.e. it cannot be expresssed exactly as a fraction and in decimal form the number goes on for ever); it crops up in many different branches of mathematics and the use of e to denote this number is sufficiently universal that it will not, normally, be defined in most papers or books. Thus, another important form for equation (2.12) is

$$y = a\,\mathrm{e}^{cx} \tag{2.15}$$

in which $\mathrm{e} \approx 2.718$. An alternative way of writing this is

$$y = a\exp(cx) \tag{2.16}$$

which means exactly the same thing as equation (2.15). (N.B. 'exp' here is a single word, it does not mean e times p. In fact, 'exp' is an abbreviation for exponential.)

Equation (2.15) is frequently the form used for modelling the variation in porosity with depth. This leads to expressions like

$$\phi = \phi_0 e^{-z/\lambda}. \tag{2.17}$$

An example is the best way to illustrate this. If the constants have the values $\phi_0 = 0.7$, $\lambda = 2\,\text{km}$, the porosity at a depth of 4 km would be

$$\begin{aligned} \phi &= 0.7 \exp(-4/2) \\ &= 0.7 \exp(-2) \\ &= 0.7 \times 0.135 \\ &= 0.0945. \end{aligned}$$

Question 2.9 What porosity would this predict for $z = 1$ km?

It is worth spending a little time considering the meaning of the constants ϕ_0 and λ. These do have fairly simple interpretations. First, remember that any number to the power zero equals one. Thus, if the depth is zero, equation (2.17) becomes

$$\begin{aligned} \phi &= \phi_0 \exp(-0/\lambda) \\ &= \phi_0 \exp(0) \\ &= \phi_0 \times 1.0 \\ &= \phi_0. \end{aligned}$$

In other words, ϕ_0 is simply the porosity at zero depth. The meaning of λ can be seen by setting z to be λ kilometres. Equation (2.17) then gives

$$\begin{aligned} \phi &= \phi_0 \exp(-\lambda/\lambda) \\ &= \phi_0 \exp(-1) \\ &= \phi_0/\text{e} \\ &\approx \phi_0/2.71 \end{aligned}$$

i.e. λ is the depth at which the porosity reduces to around one-third of its starting value.

2.8 LOGARITHMS

The logarithmic function is the final type of relationship which will be investigated in this chapter. A good place to start with logarithms is to look at the data in Table 2.6 which simply lists the result of raising 10 to the power of various integers, i.e. $10^n = 100$ if $n = 2$ etc. The logarithm of any number in

Table 2.6 Ten raised to the power of the integers between −2 and 3

n	10^n
−2	0.01
−1	0.1
0	1
1	10
2	100
3	1000

Table 2.7 Rewriting the data in Table 2.6 as a logarithmic table by swapping the columns around

Number	*Logarithm*
0.01	−2
0.1	−1
1	0
10	1
100	2
1000	3

the right-hand column is simply the exponent used in the left-hand column, e.g. the logarithm of 100 is 2, the logarithm of 1 is zero. Thus, Table 2.6 could be rewritten as a table of logarithms simply by swapping around the columns (Table 2.7).

Figure 2.11 shows part of this table plotted as a smooth curve. Once again we can see that there is nothing to stop us reading results at values other than those used to produce the plot, e.g. the logarithm of 6 is approximately 0.8 (i.e. $\log(6) = 0.8$).

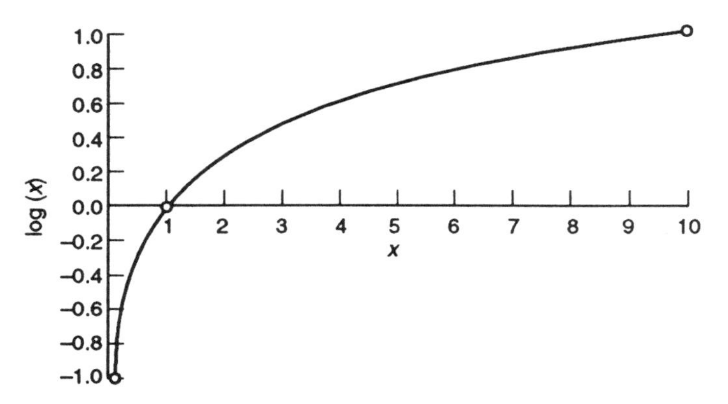

Figure 2.11 A smooth curve plotted through the points corresponding to $x = 0.1$, 1 and 10 from Table 2.7.

Another important point about Figure 2.11 is that the curve never crosses the vertical axis. If this curve was plotted for smaller values of x than 0.01 it would get even steeper so that it would never reach the $\log(x)$ axis. As a consequence, logarithms of negative numbers do not exist. If you try to use a calculator, for example to find $\log(-2)$, you will get an error message.

So, having defined the logarithm, what is it good for? There are two main uses, firstly for compressing large data ranges, secondly for reducing exponential and allometric functions to simple straight lines. The second of these will be covered in the next chapter so what about the first, i.e. compressing large ranges? Fault sizes provide a good example. Faults occur on a vast range of scales from millimetres long to hundreds of kilometres long. Now, in a particular area, the number of faults of different sizes might be something like Table 2.8 in which the fault length is tabulated against the number of faults observed of this size or larger. Note that such tables typically show that small faults are much more common than larger faults. If we attempt to plot these data on a graph, the result is as shown in Figure 2.12.

This graph is not very helpful because all the points lie on or near the axes. The problem is the large range of values that occurs; some values are very

Table 2.8 Number of faults of length greater than or equal to a given size at a particular outcrop location, e.g. there are 11 faults of length 1 m or longer

Fault length (m)	*Number*
0.001	10 109
0.01	957
0.1	132
1	11
10	1

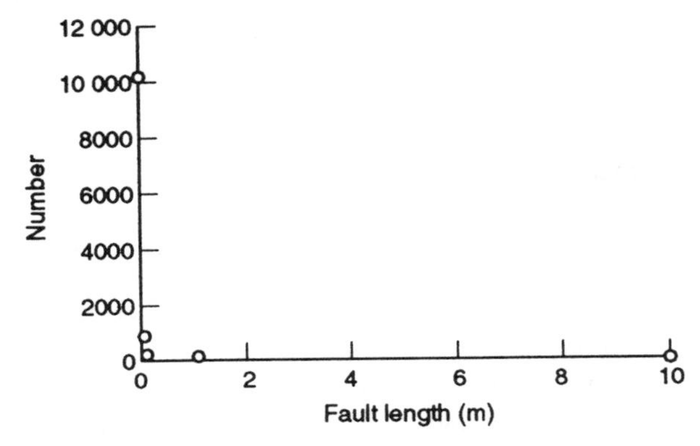

Figure 2.12 The result of plotting a graph of the data in Table 2.8.

Table 2.9 Result of taking logarithms of the columns in Table 2.8

Fault length (m)	*log(length)*	*Number*	*log(number)*
0.001	−3	10 109	4.00
0.01	−2	957	2.98
0.1	−1	132	2.12
1	0	11	1.04
10	1	1	0

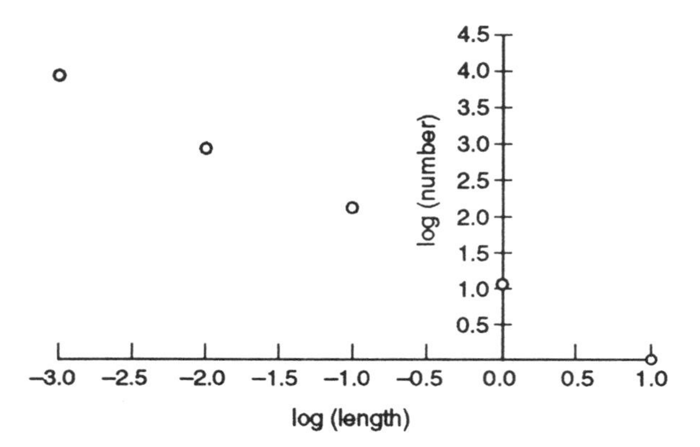

Figure 2.13 Graph of the logarithmic data from Table 2.9.

small whilst some are very large. This makes it impossible to find axes scales which enable all the data to be properly viewed. However, if we add logarithms to the columns in Table 2.8 (to give Table 2.9) and plot these instead, Figure 2.13 results. Clearly, this graph is much more informative since the data are now spread more evenly across it.

Thus, in summary, a major use for logarithms is for plotting graphs of quantities which vary over large ranges. This is something which occurs very frequently in geological problems.

2.9 LOGARITHMS TO OTHER BASES

The previous section explained how logarithms were obtained from a table showing the number 10 raised to various powers. However, it is not necessary for the number 10 to be used. If other numbers are used then the result is a logarithm in a different **base**. An alternative choice might be, for example, a base of 6. Table 2.10 shows the number 6 raised to the power of various integers and the resulting table of logarithms obtained by swapping the columns around.

Table 2.10 The result of raising 6 to the power of various integers and the resulting table of logarithms obtained by swapping the columns around

n	6^n
-2	$1/6^2 = 1/36 = 0.0278$
-1	$1/6^1 = 1/6 = 0.167$
0	$6^0 = 1$
1	$6^1 = 6$
2	$6^2 = 36$

Hence:

x	$log_6(x)$
0.0278	-2
0.167	-1
1	0
6	1
36	2

Note that, in order to indicate that these are logarithms to base 6, a 6 subscript is written after the word 'log'. Logarithms to base 10, i.e. those discussed in section 2.8, are frequently written without this subscript so that, if the subscript is missing, logs to base 10 should be understood. Logarithms to base 10 are sometimes called **common logarithms**.

Question 2.10 What number has a logarithm, in base 5, of 2 (i.e. if $\log_5(x) = 2$, what is x)? Hint: construct a table similar to Table 2.10 but for a base of 5.

A commonly used scale for quantifying sediment grain sizes is called the phi scale and this uses logarithms to base 2. The formal definition of the phi grain size is

$$\phi = -\log_2(d) \qquad (2.18)$$

where d is the grain size in millimetres. This is not as complex as it sounds as a table of base 2 logarithms shows (Table 2.11).

To convert phi values into grain sizes in millimetres it is only necessary to start at 1 mm and halve this ϕ times (e.g. for $\phi = 3$, halving 1.0 mm three times gives a grain size of $\frac{1}{8}$ mm). A convention that halving a negative number of times means doubling the same number of times must also be used (e.g. for $\phi = -3$, doubling 1.0 mm three times gives a grain size of 8 mm). However, for grain sizes which are not an integer power of 2 (0.25, 0.5, 1, 2, 4, 8 mm etc. are all integer powers of 2) this procedure will not work and the formula given by equation (2.18) must be used. Many calculators allow you to do this directly but, it you do not have access to such a calculator, there is a simple recipe for

Table 2.11 Logarithms to base 2. Note that the log increases by one for each doubling of x and $\log_2(1) = 0$

x	$log_2\ (x)$
0.25	−2
0.5	−1
1	0
2	1
4	2
8	3

converting between logarithm bases:

$$\log_b(a) = \frac{\log_c(a)}{\log_c(b)}. \tag{2.19}$$

Converting a logarithm to the base 2 into a common logarithm should make this clearer. If $b = 2$ and $c = 10$, equation (2.19) becomes

$$\begin{aligned} \log_2(a) &= \frac{\log_{10}(a)}{\log_{10}(2)} = \log_{10}(a)/0.301 \\ &= 3.32 \log_{10}(a) \end{aligned} \tag{2.20}$$

i.e. use the common logarithm and then multiply by 3.32. So, a grain size of 2.3 mm would have a phi value given by

$$\begin{aligned} \phi &= -\log_2(2.3) \\ &= -3.32 \log_{10}(2.3) \\ &= -3.32 \times 0.362 \\ &= -1.20 \end{aligned}$$

all of which leads us to a particularly common base for logarithms, namely logarithms to base e (remember $e \approx 2.718$). These are known as **natural logarithms** and are denoted either by using the subscript e (natural log of x is written $\log_e(x)$) or, more commonly, it is denoted by ln (i.e. natural log of x is written $\ln(x)$). This type of logarithm probably occurs more frequently than any other, and will be used throughout this book, so it is important to be familiar with its appearance.

FURTHER QUESTIONS

2.11 The following data were taken from the Troll 3.1 well in the Norwegian North Sea.

Depth (cm)	*Age (years)*
19.75	1 490
407.0	10 510
545.0	11 160
825.0	11 730
1158.0	12 410
1454.0	12 585
2060.0	13 445
2263.0	14 685

By plotting a graph of these data, estimate: (i) the sedimentation rate for the last 10 000 years; (ii) the sedimentation rate for the preceding 5000 years; (iii) the time since sedimentation ceased. (Data taken from Lehman, S. and Keigwin, L. (1992). Sudden changes in North Atlantic circulation during the last deglaciation. *Nature*, **356**, 757–62.)

2.12 As crystals settle out of magmas, element concentrations, C, in the remaining liquid change according to the equation

$$C = C_0 F^{(D-1)}$$

where C_0 is the concentration of the element in the liquid before crystallization began, F is the fraction of liquid remaining and D is a constant (known as the distribution coefficient). Calculate the concentration of an element after 50% crystallization (i.e. $F = 0.5$) if its initial concentration was 200 ppm and $D = 6.5$.

2.13 Radioactive minerals become less active with time according to the equation

$$\ln(a) = \ln(a_0) - \lambda t$$

where a is the radioactivity, a_0 is the initial radioactivity, t is time and λ is a constant which depends upon the mineral. If $a_0 = 1000$ counts per second and $\lambda = 10^{-7}\,\mathrm{y}^{-1}$, draw up a table and plot a graph of $\ln(a)$ against t for times ranging from 0 to 100 My. From your graph, estimate the age of a specimen which has decayed to $a = 100$ counts per second.

2.14 The variation in gravitational strength with altitude should obey the equation

$$g = g_0 + ah$$

where g is the measured strength of gravity, g_0 is the gravitational strength at sea level, a is a constant and h is height above sea level. However, the presence of metallic ore bodies, volcanic intrusions, etc. tend to increase the local strength of gravity slightly. Thus, real gravitational measurements do not quite obey this expression. Deviations from this equation can therefore be used to indicate the presence of such features. Using the figures given below, plot a graph of g against h and hence estimate g_0 and a. Hence, calculate the deviation of each measurement from its expected value. Plot a graph of this deviation as a function of position and determine the approximate extent of an ore body known to outcrop in this area.

Horizontal position, x (km)	*Altitude, h (m)*	*Gravity, g ($m\,s^{-2}$)*
0.0	150	9.809 45
0.5	100	9.809 7
1.0	170	9.809 49
1.5	200	9.809 4
2.0	150	9.809 55
2.5	130	9.809 51
3.0	120	9.809 54

2.15 The rate of accumulation, p, of carbonate sediments on a reef is given approximately by

$$p = p_0 \exp(-z/Z)$$

where p_0 and Z are constants and z is depth below sea level. (i) Calculate p at depths of 0, 2, 4, ..., 20 m if $p_0 = 3\,\mathrm{m\,ky^{-1}}$ and $Z = 20\,\mathrm{m}$. Sketch the results. (ii) Give an interpretation for the constants p_0 and Z.

Equations and how to manipulate them

3

3.1 INTRODUCTION

The last chapter introduced many of the more common mathematical functions. It is essential that you know how to manipulate expressions containing combinations of such relationships. Sometimes this will be done in order to simplify the expressions. Sometimes it will be necessary to combine several expressions to produce a new one. Very often the form of an expression is inappropriate for a particular task. Whatever the reason, this is the chapter that tells you how to go about combining, simplifying and rearranging mathematical expressions.

Some of the equations that you will see in this chapter are unfamiliar geophysical or geochemical expressions. However, these will not be derived here because this is not a geophysics or geochemistry text. Enough will be said to allow you to understand the context of the problem.

3.2 REARRANGING SIMPLE EQUATIONS

It is very obvious, but it is vitally important to appreciate, that an equation is a mathematical statement in which two expressions equal one another. Look again at the lake bed sediment example from Chapter 1:

$$\text{Age} = k \times \text{Depth}. \tag{1.1}$$

The left-hand expression is very simple, it contains 'age'. The right-hand side is also simple and is the product of k and 'Depth'. The point is that the left- and right-hand sides are stated to be equal and this is what makes (1.1) an equation. The reason that I labour this point is that the golden and unbreakable rule when manipulating equations is that, whatever you do, the left- and right-hand sides must remain equal to one another. This is simply achieved. Whenever you manipulate one side of an equation, you must perform exactly the same operation on the other side. Thus, if you add a constant to one side, you must add the same constant to the other side as well; if you double one side, you must double the other; and so on. For example, given equation (1.1), the following expressions are also true:

$$\text{Age} + 3 = (k \times \text{Depth}) + 3 \quad \text{(i.e. add 3 to both sides)}$$

$$2 \times \text{Age} = 2k \times \text{Depth} \quad \text{(i.e. double both sides)}$$

$$\sqrt{\text{Age}} = \sqrt{(k \times \text{Depth})} \quad \text{(i.e. square root both sides).}$$

By combining suitable operations on the two sides of an equation, it is possible to rearrange an equation into another form. As an example, suppose that instead of an equation which tells us the age if we know the depth (i.e. equation (1.1)) we actually need an equation which tells us the depth we would need to dig to reach sediments of a specified age. How do we do this? We must manipulate equation (1.1) to give a new equation which has 'Depth = ' on the left-hand side rather than 'Age = '. The problem is that 'Depth' only appears in combination with 'k'; it does not stand on its own. We must, somehow, remove 'k'. Now, if '$k \times$ Depth' is divided by 'k' then we are left with 'Depth'. However, if we do this to the right-hand side of equation (1.1) we must also do this to the left-hand side. This gives

$$\text{Age}/k = \text{Depth}$$

which can obviously be rewritten as

$$\text{Depth} = \text{Age}/k \tag{3.1}$$

which is the expression that we wanted.

The above example is very simple and could probably have been done almost automatically by many readers of this book. However, it is important that you read very carefully through the logic of the above example.

> Question 3.1 Manipulate equation (1.1) to give an expression for k. If, at a depth of 3 m, the age is 3000 years use your result to determine the sedimentation constant. (Assuming, of course, that equation (1.1) is valid for the lake bed in question.)

As another example, what about manipulating the slightly more complex lake sediment expression from Chapter 2

$$\text{Age} = (k \times \text{Depth}) + \text{Age of top} \tag{2.1}$$

to give an expression for depth? The problem is that there is another term on the right-hand side to remove. Should we remove k first or 'Age of top'? In fact, it does not really matter. If we try to remove 'k', by dividing by k as before, the result is

$$\text{Age}/k = \text{Depth} + (\text{Age of top}/k). \tag{3.2}$$

Note that both 'Age' and 'Age of top' now appear over k since all terms must be divided by k. An expression for the depth is then found by subtracting the second term on the right-hand side to give, after swapping left and right sides around,

$$\text{Depth} = (\text{Age}/k) - (\text{Age of top}/k). \tag{3.3}$$

Alternatively, we could have begun by attempting to remove 'Age of top' from the right-hand side of equation (2.1). Subtracting 'Age of top' from equation (2.1) gives

$$\text{Age} - \text{Age of top} = k \times \text{Depth}. \tag{3.4}$$

Dividing this by k then yields

$$\text{Depth} = (\text{Age} - \text{Age of top})/k \tag{3.5}$$

which is the same as equation (3.3) although written in a slightly different form. (If these look totally different to you, do not worry as it will be explained later in this chapter.)

Question 3.2 Starting with equation (2.1), derive an expression for the age of sediments at the surface of a dried-out lake bed. If the sedimentation constant was 5000 y m^{-1} and, at a depth of 10 m, the age was 60 000 years, determine the age of the surface sediments, this time assuming that equation (2.1) is valid.

Yet another example: how do we know the mass of the earth? The answer is that we know from the strength of gravity at the earth's surface. The strength of gravity is measured by the acceleration it causes to a falling body; a strong gravitational pull will accelerate a falling apple (say) more than a weak gravitational pull. Physicists tell us that this gravitational acceleration, g, is related to the earth's mass, M, by the equation

$$g = GM/r^2 \tag{3.6}$$

where G is a known physical constant and r is the earth's radius. To use this equation to estimate the earth's mass it must be rearranged into an expression for M. G and r^2 can be removed from the right-hand side by dividing by G and multiplying by r^2. Thus, the left-hand side must also be divided by G and multiplied by r^2 to give

$$gr^2/G = M$$

or, after swapping around

$$M = gr^2/G. \tag{3.7}$$

The values of all the symbols on the right-hand side of equation (3.7) are known. The gravitational acceleration, g, and the gravitational constant G have both been measured very accurately in physicists' laboratories, whilst the radius of the earth, r, has been known for more than 400 years (in fact some ancient Greeks had a pretty good idea too!). The values are:

$$g = 9.81\,\mathrm{m\,s^{-2}}$$
$$r = 6370\,\mathrm{km} = 6.37 \times 10^6\,\mathrm{m}$$
$$G = 6.672 \times 10^{-11}\,\mathrm{m^3\,kg^{-1}\,s^{-2}}.$$

Do not worry too much about the units of these numbers but it is important that consistent units are used (Chapter 1), so I have converted the earth's radius into metres since both g and G are given in units which include metres. Substituting these numbers into equation (3.7) gives

$$\begin{aligned} M &= 9.81 \times (6.37 \times 10^6)^2/(6.672 \times 10^{-11}) \\ &= (9.81 \times 6.37^2/6.672) \times 10^{23} \\ &= 59.7 \times 10^{23} \\ &= 5.97 \times 10^{24}\,\mathrm{kg} \end{aligned}$$

which is about six thousand million million million tons. Not bad for a small planet!

3.3 COMBINING AND SIMPLIFYING EQUATIONS

If we know the mass of the earth and can also find its volume, the earth's average density could be calculated. The volume of the earth can be estimated using the standard formula for the volume, V, of a sphere of radius r. This is

$$V = 4\pi r^3/3. \tag{3.8}$$

Using the radius of the earth given above and $\pi \approx 3.142$ gives a volume of

$$\begin{aligned} V &= 4 \times 3.142 \times (6.37 \times 10^6)^3/3 \\ &= (4 \times 3.142 \times 6.37^3/3) \times 10^{18} \\ &= 1083 \times 10^{18} \\ &= 1.083 \times 10^{21}\,\mathrm{m^3}. \end{aligned}$$

The density (usually denoted by rho, ρ) is related to mass and volume by

$$\rho = M/V. \tag{3.9}$$

Thus, using the mass and volume already found, the average density of the earth is given by

$$\begin{aligned}\rho &= (5.97 \times 10^{24})/(1.083 \times 10^{21}) \\ &= (5.97/1.08) \times 10^3 \\ &= 5.51 \times 10^3 \\ &= 5510\,\mathrm{kg\,m^{-3}}\end{aligned}$$

which is more than five times the density of water (which is around $1000\,\mathrm{kg\,m^{-3}}$).

The above derivation was a little tortuous. It is possible to combine the three expressions (equations (3.7)–(3.9)) into a single expression for density. This means less numerical calculation is necessary and fewer errors will be made. Equation (3.9) tells us to divide mass by volume to give density. The mass is given by equation (3.7) whilst the volume is given by equation (3.8). Therefore we can immediately write down

$$\begin{aligned}\rho &= M/V = \text{equation (3.7)/equation (3.8)} \\ &= \frac{gr^2/G}{4\pi r^3/3}.\end{aligned} \tag{3.10}$$

In other words, it is always possible to replace an expression (e.g. M) by another equal expression (in this case gr^2/G). Since the initial and replacement expressions are equal, the right-hand side of the equation does not change its value and the equation remains true. We now have an expression for density which could be evaluated by substituting the known values for g, r and G. However, equation (3.10) looks a bit daunting. Evaluating it is not any easier than separately evaluating M and V as before. Fortunately, it is possible to simplify. First, we multiply both the top of the right-hand side and the bottom of the right-hand side by G. Since we are multiplying the whole expression then by G/G (which equals 1.0), this has no effect upon the left-hand side. The result is

$$\begin{aligned}\rho &= \frac{G(gr^2/G)}{G(4\pi r^3/3)} \\ &= \frac{gr^2}{4G\pi r^3/3}\end{aligned} \tag{3.11}$$

since the two Gs on the top cancel each other. Now we can do a similar trick to remove the division by 3. Multiplying top and bottom by 3 yields

$$\rho = \frac{3gr^2}{4G\pi r^3}. \tag{3.12}$$

Finally, by noting that $r^3 = r^2 r$ and cancelling, this expression can be further reduced to

$$\rho = \frac{3gr^2}{4G\pi r^2 r}$$

$$= 3g/(4G\pi r). \qquad (3.13)$$

This is much simpler to use than equation (3.10). Let us just check that it gives the right result:

$$\begin{aligned}\rho &= 3g/(4G\pi r)\\ &= 3 \times 9.81/(4 \times 6.672 \times 10^{-11} \times 3.142 \times 6.37 \times 10^6)\\ &= [3 \times 9.81/(4 \times 6.672 \times 3.142 \times 6.37)] \times 10^5\\ &= 0.0551 \times 10^5\\ &= 5510\,\mathrm{kg\,m^{-3}}\end{aligned}$$

as before. Note that this result is about twice the density of typical rocks found at the earth's surface. We can therefore conclude that the deep earth must be much denser than the near surface for the average to come out so high.

Question 3.3 Prove that if

$$w = 3y/(4z)$$

and

$$x = 2y/(4z)$$

then

$$w/x = 1.5.$$

These then, are the basic tools for equation manipulation: (i) you can add, multiply, divide, double, halve, subtract or perform any other operation you like, provided that you do exactly the same to both sides of an equation; (ii) you can always replace an expression by any other expression which is equal to it.

3.4 MANIPULATING EXPRESSIONS CONTAINING BRACKETS

An important mathematical skill is the ability to use brackets effectively. Sometimes an expression can be made a great deal easier to understand, and easier to use, if brackets are added or removed. Brackets will normally be added into an equation by a procedure called **factorization**, whilst the reverse operation which removes brackets is achieved by multiplying out.

It is probably easiest to begin by explaining how to multiply out brackets. The following example may seem extremely trivial at first. However, despite its apparent simplicity it actually contains all the ideas necessary for carrying

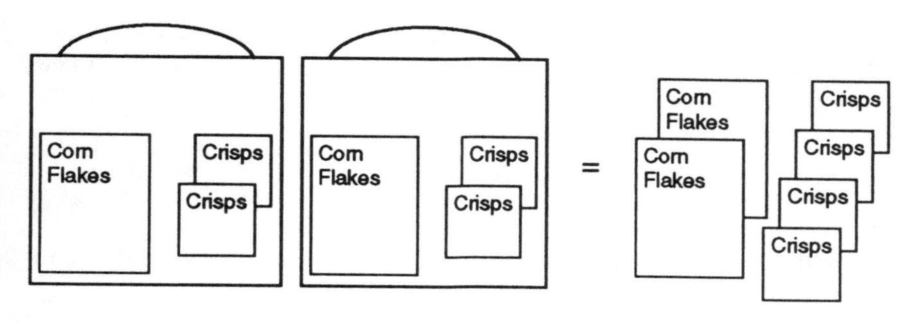

Figure 3.1 Two shopping bags each containing one packet of cereal and two packets of crisps hold two packets of cereal and four of crisps altogether. This is exactly the same as the algebraic problem below. Two lots of $x + 2y$ gives $2x$ and $4y$.

out algebraic manipulations of this type. Imagine two shopping bags each containing a box of cereal and two packets of crisps. How many boxes of cereal and packets of crisps is that altogether? Obviously it is two boxes of cereal and four packets of crisps (Figure 3.1).

Multiplying out brackets is exactly the same. All you do is multiply the items inside the brackets (equivalent to a bag of items) by the number outside the brackets (equivalent to the number of bags). Thus

$$2(x + 2y) = 2x + 4y \qquad (3.14)$$

is exactly the same as the shopping bag problem. The left-hand side says that there are two lots of $(x + 2y)$ and the right-hand side says that this is the same as two x and four y. If you think of x as a box of cereal and y as a packet of crisps you should see the equivalence of the two problems. In practice, all you do is multiply each of the terms inside the bracket by the number outside. This rather easy example actually contains all the mathematics you need to tackle any other cases. For example,

$$2.3(x + 2y + 4z) = 2.3x + 4.6y + 9.2z \qquad (3.15)$$

is calculated exactly the same way as before. Simply multiply each of the terms inside the bracket by the number outside. If the number outside is a symbol rather than a number this leads to examples such as

$$a(x + 2y + 4z) = ax + 2ay + 4az \qquad (3.16)$$

which is no different from the earlier examples; simply multiply each term in the brackets by a. A final difficulty is if the number outside the bracket is a more complex expression such as $(a + 3)$ giving

$$(a + 3)(x + 2y + 4z) = (a + 3)x + 2(a + 3)y + 4(a + 3)z. \qquad (3.17)$$

As you can see, this is still done the same way. However, this time we can take things a stage further. Each of the resulting terms in equation (3.17) is a new problem in multiplying out brackets. For example, the first term on the right-hand side is

$$\begin{aligned}(a+3)x &= x(a+3)\\ &= ax+3x.\end{aligned} \tag{3.18}$$

The other terms in equation (3.17) can be similarly multiplied out leading to a final answer of

$$(a+3)x+2(a+3)y+4(a+3)z = ax+3x+2ay+6y+4az+12z \tag{3.19}$$

Question 3.4 Multiply out the brackets in the following examples.

(i) $5(x+2y)$, (ii) $5(x+2.2y)$, (iii) $5.5(x+2y)$, (iv) $5a(x+2y)$, (v) $(x-2y)(x+2y)$, (vi) $(x+2y)^2$.

Question 3.5 Earlier in this chapter I stated that the expression

$$\text{Depth} = (\text{Age}/k) - (\text{Age of top}/k) \tag{3.3}$$

was exactly the same as

$$\text{Depth} = (\text{Age} - \text{Age of top})/k. \tag{3.5}$$

Verify this by rewriting equation (3.5) in the form

$$\text{Depth} = (1/k)(\text{Age} - \text{Age of top})$$

and multiplying out the bracket.

Factorization is the reverse process to multiplying out of brackets. For example, equation (3.18) above was

$$x(a+3) = ax+3x. \tag{3.18}$$

Factorization is the process of writing this the other way around:

$$ax+3x = x(a+3). \tag{3.20}$$

Its main use is for simplifying the appearance of more complex expressions and relies upon spotting **common factors.** The trick is to spot that both terms on the left-hand side of equation (3.20) contain a factor x, i.e. they are both equal to some quantity multiplied by x (a times x for the first term and three times x for the second). This x can be taken out as a common factor leaving $a+3$ inside the bracket. Another, more difficult, example might be

$$3.2xy+6.4xw+z = ? \tag{3.21}$$

The first two terms have a common factor of $3.2x$ which can therefore be written outside a bracket containing $y + 2w$. Thus, the solution is

$$3.2xy + 6.4xw + z = 3.2x(y + 2w) + z. \tag{3.22}$$

Note that the third term does not have any factors in common with the first two and is therefore left alone.

Question 3.6 Factorize $6ax + 3ay$.

Factorization can be used to derive an equation for the density of a wet, porous sandstone. This rock will be partly made from sand grains with a density of ρ_s and partly made from water with a density of ρ_w. Hence, the average density of the sample will be somewhere between ρ_s and ρ_w. If the porosity, ϕ, is low the density will be close to that of the sand grains but if the porosity is higher then the average density will be a little closer to that of water. More mathematically, take a specimen of this sandstone which has a volume V and mass m. This mass will be made up from the mass of water in the volume plus the mass of the grains, i.e.

$$m = m_w + m_s \tag{3.23}$$

where m_w and m_s are the masses of water and sand respectively. However, the mass of water is given by the product of the volume of water and the density of water whilst m_s is similarly given by the grain density times the volume of grains. Thus equation (3.23) can be written

$$m = V_w\rho_w + V_s\rho_s \tag{3.24}$$

where V_w and V_s are the volumes of water and sand in the sample. The volume of water is equal to the volume of the sample multiplied by the porosity (e.g. porosity equal to 0.5 implies that half the total volume is water, a porosity of 0.25 implies one-quarter of the total volume is water). Thus

$$V_w = \phi V. \tag{3.25}$$

The remaining volume must be sand and therefore

$$V_s = V - \phi V. \tag{3.26}$$

Substituting these volumes into equation (3.24) gives

$$m = \phi V\rho_w + (V - \phi V)\rho_s. \tag{3.27}$$

Multiplying out the bracket then leads to

$$m = \phi V\rho_w + V\rho_s - \phi V\rho_s. \tag{3.28}$$

Now V is common to all the terms on the right-hand side and can therefore be taken out as a common factor to give

$$m = V(\phi\rho_w + \rho_s - \phi\rho_s). \tag{3.29}$$

Finally, dividing by V and performing a further factorization gives the required average density

$$\begin{aligned}\rho &= m/V \\ &= \phi\rho_w + \rho_s - \phi\rho_s \\ &= \phi\rho_w + (1-\phi)\rho_s. \end{aligned} \tag{3.30}$$

Question 3.7 Using equation (3.30), plot a graph of how the density varies as porosity changes from zero to one. Assume

$$\rho_w = 1000\,\text{kg m}^{-3}$$
$$\rho_s = 2500\,\text{kg m}^{-3}$$

3.5 'REARRANGING' QUADRATIC EQUATIONS

Chapter 2 introduced a quadratic expression for calculating temperature for the deeper parts of the earth. This was in the form

$$\text{Temperature} = (-8.255 \times 10^{-5})z^2 + 1.05z + 1110. \tag{2.6}$$

How can this be rearranged to allow calculation of the depth for a given temperature, e.g. at what depth is the temperature 2000 °C? In fact, such a rearrangement is rather difficult. To solve this problem it is first necessary to discuss a technique called finding the **roots** of a quadratic equation.

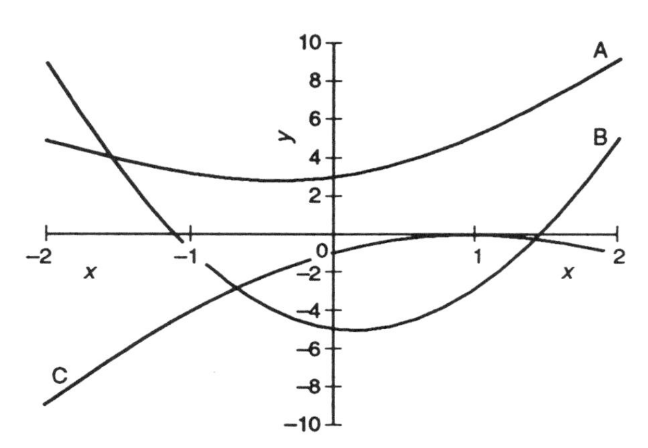

Figure 3.2 The roots of a quadratic equation are the points where the quadratic curve crosses the horizontal axis, i.e. the points where the quadratic function equals zero. Thus: curve A has no roots; curve B has two roots at about $x = -1.1$ and $x = 1.5$; curve C has one root somewhere near $x = 1$.

The roots of a quadratic equation are the values of the variable which make the quadratic expression equal zero. Figure 3.2 should help. The most general way to find these roots is to use the following method. The roots of the quadratic equation

$$y = ax^2 + bx + c \tag{3.31}$$

are the values of x for which

$$y = 0 \tag{3.32}$$

or, equivalently,

$$0 = ax^2 + bx + c. \tag{3.33}$$

These are given by

$$x = \frac{-b \pm \sqrt{(b^2 - 4ac)}}{2a} \tag{3.34}$$

where $\pm$ means 'either add or subtract'. For example, curve B in Figure 3.2 had the equation

$$y = 3x^2 - x - 5 \tag{3.35}$$

i.e.

$$a = 3, \qquad b = -1 \qquad c = -5.$$

Note that both b and c are negative in this example. Substituting these values into equation (3.34) gives

$$\begin{aligned} x &= [1 \pm \sqrt{(1 + 4 \times 3 \times 5)}]/[2 \times 3] \\ &= [1 \pm \sqrt{61}]/6 \\ &= [1 + \sqrt{61}]/6 \quad \text{or} \quad = [1 - \sqrt{61}]/6 \\ &= 1.47 \qquad\qquad \text{or} \quad = -1.14 \end{aligned} \tag{3.36}$$

which are indeed the points where curve B crosses the horizontal axis in Figure 3.2. If, on the other hand, we look at curve A, this has the equation

$$y = x^2 + x + 3 \tag{3.37}$$

i.e.

$$a = 1 \qquad b = 1 \qquad c = 3.$$

Substituting these values into equation (3.34) gives

$$\begin{aligned} x &= [-1 \pm \sqrt{(1 - 12)}]/2 \\ &= [-1 \pm \sqrt{-11}]/2 \end{aligned} \tag{3.38}$$

but $\sqrt{-11}$ has no solution (because you cannot find the square root of a negative number) and therefore there are no roots (which is what Figure 3.2 shows).

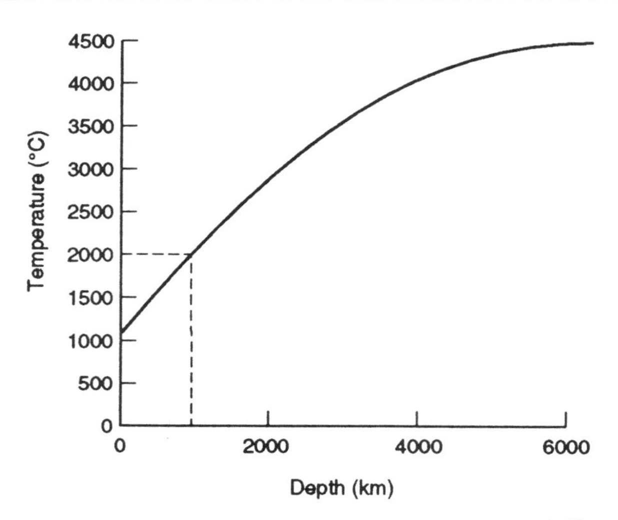

Figure 3.3 At what depth is the temperature 2000 °C?

Question 3.8 Find the roots for curve C where

$$y = -x^2 + 2x - 1.$$

We can now attempt the problem of how to find the depth for a particular temperature. Take the case where we wish to know the depth at which the temperature is 2000 °C (Figure 3.3). Substituting this temperature into equation (2.6) gives

$$2000 = (-8.255 \times 10^{-5})z^2 + 1.05z + 1110 \tag{3.39}$$

which is not quite in the form that we need. We have a method for solving equations like (3.33) where the quadratic expression equals zero. Here we have an equation which equals 2000. However, this is simply remedied by subtracting 2000 from both sides to give

$$0 = (-8.255 \times 10^{-5})z^2 + 1.05z - 890 \tag{3.40}$$

which is equivalent to equation (3.33) with values for a, b and c of

$$a = -8.255 \times 10^{-5} \qquad b = 1.05 \qquad c = -890.$$

Substituting these into equation (3.34) (and remembering that z is the variable in equation (2.6) not x) gives

$$z = \frac{-1.05 \pm \sqrt{[1.05^2 - (4 \times 8.255 \times 10^{-5} \times 890)]}}{-2 \times 8.255 \times 10^{-5}} \tag{3.41}$$

which has solutions of $z = 913$ km and $z = 11\,806$ km. There are two depths because at a depth of 11 806 km you are about 913 km from the surface on the far side of the earth.

Question 3.9 Find the depths at which temperature reaches 3000 °C.

FURTHER QUESTIONS

3.10 The density, ρ, of an air-filled porous rock is given by

$$\rho = \rho_g[1 - (V_p/V)]$$

where ρ_g is the density of the grains making up the rock, V_p is the volume occupied by pore space and V is the total volume. By combining this with equation (3.9) prove that the grain density is given by

$$\rho_g = M/(V - V_p)$$

where M is the mass of the rock sample. Hence, calculate both the average density and the grain density of a sample with a volume of 0.11 m^3, a mass of 205 kg and a porosity of 0.32.

3.11 Stokes' law states that the velocity at which a spherical particle suspended in a fluid settles to the bottom is given by

$$v = \frac{2(\rho_p - \rho_f)gr^2}{9\eta}$$

where v is the velocity of descent, ρ_p and ρ_f are the densities of particle and fluid respectively, g is the acceleration due to gravity, r is the particle radius and η is a property of the fluid known as its **viscosity**. Assuming that grains of different sizes have identical densities, show that the ratio of the settling velocities for two different grain sizes is

$$\frac{v_1}{v_2} = \left(\frac{r_1}{r_2}\right)^2$$

where v_1 and v_2 are the velocities for grains of radius r_1 and r_2 respectively. If a grain of radius 0.1 mm, suspended in a lake, takes 10 days to settle to the lake bottom, how long would it take a grain of radius 1 mm?

3.12 (i) Rearrange

$$0 = ax^2 + bx + c$$

into an equation for b.

(ii) Use your answer from (i) to verify that

$$b^2 - 4ac = a^2x^2 + (c^2/x^2) - 2ac.$$

(iii) Verify that the answer to (ii) could be factorized to yield

$$b^2 - 4ac = [ax - (c/x)]^2.$$

(Hint: it is easiest to do this by multiplying out the above expression.)

(iv) Use the above answers to verify that one of the roots of a quadratic expression is given by

$$x = \frac{-b + \sqrt{(b^2 - 4ac)}}{2a}.$$

More advanced equation manipulation

4

4.1 INTRODUCTION

In the last chapter you were introduced to methods for manipulating simple equations. In this chapter we will look at a few, more advanced, techniques for equation manipulation. In particular, I shall discuss manipulation of equations containing exponentials and logarithms. I will also look at the topic of simultaneous equations in which several equations must be manipulated at the same time in order to solve a problem.

This chapter also introduces techniques for checking equations for errors. They may have been wrong on the first place (e.g. due to a printing error) or you may have made mistakes during your manipulations. Either way, it is useful to be able to check that equations are reasonable.

4.2 EXPRESSIONS INVOLVING EXPONENTIALS AND LOGARITHMS

In Chapter 2 we looked at expressions involving exponentials such as

$$\phi = \phi_0 \, e^{-z/\lambda} \qquad (2.17)$$

for variation in porosity, ϕ, with depth, z. We also looked at logarithms and it was stated that these could be used to recast the graph of an exponential expression into the form of a straight line. This procedure is essential if, for example, you wish to rewrite equation (2.17) to give the depth at which a particular porosity occurs. A little further revision on the properties of logarithms is first needed. The point to remember about logarithms is that they are simply the reverse operation to raising to a power (section 2.8). From this, it follows that

$$\log_y(y^x) = x \tag{4.1}$$

where y is any base for the logarithm.

For example, $10^2 = 100$ and $\log(100) = 2$. i.e. $\log(10^2) = 2$. Another useful result is that

$$\log(ab) = \log(a) + \log(b) \tag{4.2}$$

where again the log can be to any base. In other words, the logarithm of any two numbers multiplied together is equal to the sum of the logarithms of the numbers. For example,

$$\log(12) = \log(3) + \log(4) \quad (\text{since } 3 \times 4 = 12)$$

and also

$$\log(12) = \log(6) + \log(2) \quad (\text{since } 2 \times 6 = 12)$$

or even

$$\log(12) = \log(10) + \log(1.2) \quad (\text{since } 10 \times 1.2 = 12).$$

Equation (4.2) can be generalized into

$$\log(abc\ldots) = \log(a) + \log(b) + \log(c) + \cdots \tag{4.3}$$

i.e. the logarithm of a series of numbers multiplied together equals the sum of their logarithms. For example,

$$\ln(2 \times 3 \times 4.712 \times f) = \ln(2) + \ln(3) + \ln(4.712) + \ln(f).$$

A special case of equation (4.3) is the logarithm of a number raised to a power. In this case we get

$$\begin{aligned} \log(x^n) &= \log(x) + \log(x) + \log(x) + \cdots \quad \text{i.e. } n \text{ terms like this} \\ &= n\log(x). \end{aligned} \tag{4.4}$$

Finally, substraction of two logarithms is equivalent to division of the **arguments** (the argument of $\log(b)$ is b, the argument of $\ln(f)$ is f and so on). Thus

$$\log(a/b) = \log(a) - \log(b) \tag{4.5}$$

and

$$\ln(a/b) = \ln(a) - \ln(b). \tag{4.6}$$

These rules allow the following transformation of equations such as (2.17). Starting with

$$\phi = \phi_0 \mathrm{e}^{-z/\lambda} \tag{2.17}$$

and taking the natural logarithm of both sides gives

$$\ln(\phi) = \ln(\phi_0 \mathrm{e}^{-z/\lambda}). \tag{4.7}$$

Now, using the rule about logarithms of products (equation (4.3)), leads to

$$\ln(\phi) = \ln(\phi_0) + \ln(e^{-z/\lambda}). \tag{4.8}$$

Finally, equation (4.1) produces

$$\ln(\phi) = \ln(\phi_0) - z/\lambda \tag{4.9}$$

which implies that a graph of $\ln(\phi)$ against z is a straight line of gradient $(-1/\lambda)$ and intercept $\ln(\phi_0)$ (Figure 4.1).

We now have an expression which can be rearranged for depth. First add z/λ to both sides of equation (4.9) to give

$$z/\lambda + \ln(\phi) = \ln(\phi_0). \tag{4.10}$$

Then subtract $\ln(\phi)$

$$z/\lambda = \ln(\phi_0) - \ln(\phi). \tag{4.11}$$

Finally, multiply by λ

$$z = \lambda[\ln(\phi_0) - \ln(\phi)]. \tag{4.12}$$

Alternatively (using equation (4.6)) this can be expressed as

$$z = \lambda \ln(\phi_0/\phi). \tag{4.13}$$

Thus, equation (4.12) or (4.13) can now be used to obtain the depth at which a specific porosity occurs. For example, for $\lambda = 2\,\text{km}$ and $\phi_0 = 0.7$ a porosity of 0.35 would occur at a depth of

$$\begin{aligned} z &= 2\ln(0.7/0.35) \\ &= 2\ln(2) \\ &= 2 \times 0.693 \\ &= 1.39\,\text{km}. \end{aligned}$$

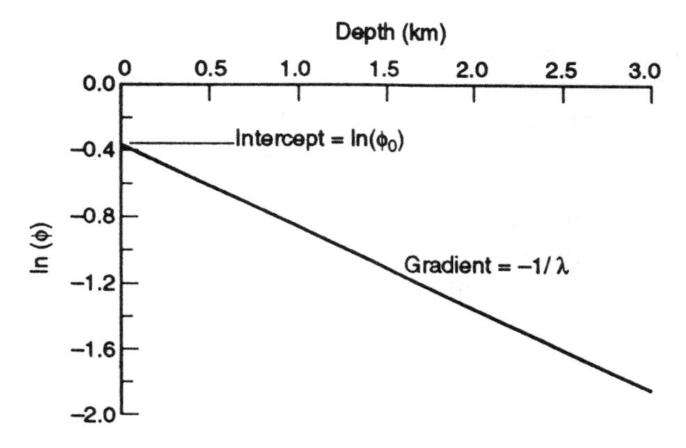

Figure 4.1 Graph of ln(porosity) against depth assuming a porosity–depth relationship in the form of equation (2.17) and for an initial porosity of 0.7 and $\lambda = 2\,\text{km}$.

A geochemical example may serve to reinforce these ideas. Radioactive dating is based upon the fact that the amount of a radioactive material decreases with time. Thus, if you know how much of the radioactive substance is present now and how much was present when a mineral was formed, it is possible to calculate the time since formation. The amount of material present is given by

$$Q = Q_0 \exp(-t/\lambda) \tag{4.14}$$

where Q is the amount, Q_0 is the original amount, t is time since formation and λ is a constant which depends upon the radioactive element. This equation is in the same form as the porosity versus depth expression discussed above. It is possible therefore to rearrange it in exactly the same way to give

$$t = \lambda \ln(Q_0/Q) \tag{4.15}$$

which can then be used to calculate the specimen's age. In practice, radioactive dating is usually more complicated than this but the above analysis nevertheless contains the essence of such techniques.

Question 4.1 If $Q_0 = 100$ ppb (parts per billion), $Q = 5$ ppb and $\lambda = 5 \times 10^6$ years, calculate t.
Question 4.2 If the age of a specimen is known to be 100 My and the present concentration of ^{235}U is 10 ppb, find the amount of uranium originally present in the sample. (Use $\lambda = 1.01 \times 10^9$ years for ^{235}U.)

4.3 SIMULTANEOUS EQUATIONS

Perhaps the best way to illustrate **linear simultaneous equations** is to launch straight into a problem. In Chapter 2, I introduced quadratic and polynomial expressions for approximating the temperature versus depth function of the earth. How were these expressions derived?

To keep this example as simple as possible, I will calculate temperature as a function of distance, r, from the earth's centre. This is slightly easier than calculating temperature as a function of depth from the surface. The starting point is to observe that the general shape of the temperature versus depth function is symmetric about the earth's centre with a steep gradient and low temperature near the earth's surface (i.e. at large r) and a zero gradient and high temperature at its centre (i.e. at $r = 0$). Now, from the discussion of quadratic equations given in Chapter 2, it should be clear that the expression

$$T = ar^2 + b \tag{4.16}$$

can be made to have these properties. In this expression, a and b are constants and, in order to have the right properties, b should be the temperature at the earth's centre and a should be negative (so that temperature reduces as r

increases). The problem then becomes simply one of determining values for a and b which enable equation (4.16) to be a good approximation to the true temperature profile. The next step is to choose two values for r where the temperature is known. In this example I will use $T = 4300\,°C$ at $r = 1260$ km and $T = 1150\,°C$ at $r = 6260$ km (from Table 2.3 after converting from depth to distance from earth's centre). If these are substituted into equation (4.16), the results are

$$4300 = 1\,587\,600a + b \tag{4.17}$$

and

$$1150 = 39\,187\,600a + b. \tag{4.18}$$

This is the **system** of linear simultaneous equations that we have to solve. The point here is that we have two different equations each of which contains the same unknowns (i.e. a and b). Now, whenever we have two such equations it is usually possible to combine them to deduce the two unknowns. More generally, whenever we have the same number of unknowns as we have different equations, the equations can usually be combined to calculate the unknown values. In the case of equations (4.17) and (4.18), these may be solved as follows:

1. Rearranging equation (4.17) to give an expression for b gives

$$b = 4300 - 1\,587\,600a. \tag{4.19}$$

2. Substituting this result into equation (4.18) gives

$$1150 = 39\,187\,600a + 4300 - 1\,587\,600a. \tag{4.20}$$

3. Rearranging to give an expression for a gives

$$\begin{aligned} a &= (1150 - 4300)/(39\,187\,600 - 1\,587\,600) \\ &= -3150/37\,600\,000 \\ &= -8.378 \times 10^{-5}. \end{aligned} \tag{4.21}$$

4. Finally, this result may be substituted back into equation (4.19) to give

$$\begin{aligned} b &= 4300 - (1\,587\,600 \times -8.378 \times 10^{-5}) \\ &= 4300 + 133 \\ &= 4433. \end{aligned} \tag{4.22}$$

In some cases, there are short cuts which can be made in this procedure. However, the method given above will always produce a solution if one exists.

Sometimes a solution does not exist. This occurs if the equations are not **linearly independent**. For example, the equations

$$2 = 2a + 3b \tag{4.23}$$

and

$$4 = 4a + 6b \tag{4.24}$$

are not linearly independant since the second equation is simply the first equation after doubling. Another example would be

$$1 = 2a + 3b \tag{4.25}$$

and

$$2 = 2a + 3b \tag{4.26}$$

since these equations cannot both be true simultaneously ($2a + 3b$ cannot equal both 1 and 2). Having obtained a solution (e.g. $a = -8.378 \times 10^{-5}$ and $b = 4433$ in equations (4.17) and (4.18)), it is a good idea to test the solution by substituting the results back into the original equations. Thus, using the example given above, equations (4.17) and (4.18) become

$$\begin{aligned} 4300 &= (1\,587\,600 \times -8.378 \times 10^{-5}) + 4433 \\ &= -133 + 4433 = 4300 \end{aligned} \tag{4.27}$$

and

$$\begin{aligned} 1150 &= (39\,187\,600 \times -8.378 \times 10^{-5}) + 4433 \\ &= -3283 + 4433 = 1150 \end{aligned} \tag{4.28}$$

The above example was for the case of two equations and two unknowns. In cases where there are more unknowns (remember: the number of equations and the number of unknowns must be the same), the procedure is very similar. As an example, take the following case which has three equations and three unknowns:

$$5 = 2x + 4y + 2z \tag{4.29}$$

$$10 = x + y - z \tag{4.30}$$

$$1 = 2x - 2y + 3z \tag{4.31}$$

where x, y and z are the unknowns to be determined. Thus:

Step (i) Rearrange any one of the expressions to give one unknown in terms of the other two. It does not matter which equation and which unknown are used. For example, equation (4.30) gives

$$z = x + y - 10. \tag{4.32}$$

Step (ii) Substitute into the remaining equations to give a new system of equations. Thus, substituting equation (4.32) into equation (4.29) yields

$$\begin{aligned} 5 &= 2x + 4y + 2(x + y - 10) \\ &= 2x + 4y + 2x + 2y - 20 \\ &= 4x + 6y - 20 \end{aligned}$$

giving

$$25 = 4x + 6y. \tag{4.33}$$

Similarly, substituting equation (4.32) into equation (4.31) gives

$$\begin{aligned}1 &= 2x - 2y + 3(x + y - 10) \\ &= 2x - 2y + 3x + 3y - 30 \\ &= 5x + y - 30\end{aligned}$$

i.e.

$$31 = 5x + y. \tag{4.34}$$

Step (iii) Repeat previous steps for the new, smaller, system of equations. In this case we have a new system consisting of equations (4.33) and (4.34), i.e. two equations and two unknowns. Thus, we may now solve as for the earlier problem. Rearranging equation (4.34) leads to

$$y = 31 - 5x \tag{4.35}$$

which, upon substitution into equation (4.33), gives

$$\begin{aligned}25 &= 4x + 6(31 - 5x) \\ &= 4x + 186 - 30x \\ &= -26x + 186\end{aligned}$$

giving

$$-26x = -161$$

i.e.

$$x = 161/26 = 6.1923. \tag{4.36}$$

Step (iv) Back substitution. Having obtained one unknown, substitute into earlier equations to determine the others. First, substitute equation (4.36) into equation (4.35) to give

$$\begin{aligned}y &= 31 - (5 \times 6.1923) \\ &= 31 - 30.9615 = 0.0385.\end{aligned} \tag{4.37}$$

Finally, substitute both equations (4.36) and (4.37) into equation (4.32) to give

$$\begin{aligned}z &= 6.1923 + 0.0385 - 10 \\ &= -3.7692.\end{aligned} \tag{4.38}$$

Substituting these back into the original equations (i.e. equations (4.29), (4.30) and (4.31)) gives

$$5 = 12.3846 + 0.154 - 7.5384 = 5.002 \tag{4.39}$$

$$10 = 6.1923 + 0.0385 + 3.7692 = 10 \tag{4.40}$$

$$1 = 12.3846 - 0.077 - 11.3076 = 1 \tag{4.41}$$

respectively. Note that small rounding errors have produced a slight discrepancy in expression (4.39).

Question 4.3 Solve the following system of equations for a, b and c:

$$\begin{aligned} 10.3 &= 3a + 2b + c \\ 7 &= a + b + 1.3c \\ 5 &= 10a - 1.35b - 1.1c. \end{aligned}$$

Check your answer by substituting your results back into the above equations.

4.4 QUALITY ASSURANCE

It is a good idea to consider ways that answers can be checked. It is so easy to make mistakes that methods for checking answers are immensely valuable. This section suggests three approaches and any one of them may, or may not, be useful for checking your mathematics in a particular case. The methods are:

1. *Approximation*: does the answer seem to be about the right size?
2. *Dimensional analysis*: does the answer have the right units?
3. *Special cases*: does the answer give sensible results in simple cases?

Approximation is done by simplifying the numbers in a problem so that calculations can be performed quickly and without error. In the porosity versus depth problem where

$$\phi = \phi_0 \mathrm{e}^{-z/\lambda} \tag{2.17}$$

we might have the case where $z = 4.2\,\mathrm{km}$, $\lambda = 1.9\,\mathrm{km}$ and $\phi_0 = 0.92$. The porosity is then given by

$$\begin{aligned} \phi &= 0.92\,\mathrm{e}^{-4.2/1.9} \\ &= 0.26. \end{aligned}$$

To check this, make the approximations $z = 4\,\mathrm{km}$, $\lambda = 2\,\mathrm{km}$, $\mathrm{e} \approx 3$ and $\phi_0 \approx 0.9$. This produces

$$\begin{aligned} \phi &= 0.9 \times 3^{-2} \\ &= 0.9/3^2 = 0.9/9 \\ &= 0.1 \end{aligned}$$

which is highly reliable since it does not require the error-prone use of a calculator. However, this is not very close to the 'exact' answer of 0.26 given above. Careful checking of the arithmetic shows an error was made in the 'exact' calculation. (In fact $z = 2.4\,\mathrm{km}$ was used instead of $z = 4.2\,\mathrm{km}$. Note that this type of error is very easy to make on a calculator.) Recalculation using the correct value for z yields a porosity of 0.101 which is much closer to the estimated value.

In fact, this agreement is unusually good since some of the values (λ and e) were increased when approximated whilst the others (z and ϕ) were decreased. It happens that, in this case, these alterations very nearly cancel one another exactly. This will not usually happen quite so well! The closeness of the approximation also depends strongly upon the function involved. Expressions involving exponentials generally approximate rather poorly whilst other functions (particularly straight lines for example) can give very good answers after approximation.

Question 4.4 Using approximation, check your answer to question 3.9 from the last chapter. (Hint: use $a \approx -1 \times 10^{-4}$, $b \approx 1$, and approximate c as well. Do as much as possible without the use of a calculator.)

Dimensional analysis is a method for finding the correct units for an answer. The reason that this can be used to check mathematics is that, frequently, the units are known anyway and the result from a dimensional analysis should agree with the expected units. Thus, dimensional analysis allows you to check the form of an equation rather than the accuracy of a numerical answer (cf. approximation). Take the earth's average density example from Chapter 2. This was

$$\rho = 3g/4G\pi r \tag{3.13}$$

where the numerical values and units for the symbols involved are

$$g = 9.81\ \mathrm{m\,s^{-2}}$$
$$r = 6370\ \mathrm{km} = 6.37 \times 10^{6}\ \mathrm{m}$$
$$G = 6.672 \times 10^{-11}\ \mathrm{m^3\,kg^{-1}\,s^{-2}}.$$

The result, which is a density, should be in units of $\mathrm{kg\,m^{-3}}$. What are the units predicted by equation (3.13)? The procedure is to re-evaluate equation (3.13) using the units of the symbols rather than their numerical values. Thus

$$\text{Units of } \rho = \frac{\text{Units of } 3 \times \text{Units of } g}{\text{Units of } 4 \times \text{Units of } G \times \text{Units of } \pi \times \text{Units of } r}. \tag{4.42}$$

Numbers such as 3, 4 or π do not normally have units. These are called **dimensionless**. The units of the other terms are:

$$\text{Units of } g = \mathrm{m\,s^{-2}}$$
$$\text{Units of } G = \mathrm{m^3\,kg^{-1}\,s^{-2}}$$
$$\text{Units of } r = \mathrm{m}.$$

Substituting these into equation (4.42) gives

$$\text{Units of } \rho = \mathrm{m\,s^{-2}}/[(\mathrm{m^3\,kg^{-1}\,s^{-2}})(\mathrm{m})].$$

Multiplying the m^3 on the bottom line by the m on the bottom line then gives

$$\text{Units of } \rho = \text{m s}^{-2}/(\text{m}^4\,\text{kg}^{-1}\,\text{s}^{-2})$$

Finally, cancelling m s^{-2} from top and bottom yields

$$\begin{aligned}\text{Units of } \rho &= 1/(\text{m}^3\,\text{kg}^{-1})\\ &= \text{kg m}^{-3}\end{aligned}$$

as expected. Thus, equation (3.13) is dimensionally balanced. Equations which do not balance dimensionally must be wrong. Unfortunately, equations are not necessarily correct just because they balance; it is possible for an incorrect equation to balance perfectly.

Question 4.5 Rearrange

$$g = GM/r^2 \tag{3.6}$$

into an equation for G. Hence, using the fact that

$$\begin{aligned}\text{Units of } g &= \text{m s}^{-2}\\ \text{Units of } M &= \text{kg}\\ \text{Units of } r &= \text{m}\end{aligned}$$

show that

$$\text{Units of } G = \text{m}^3\,\text{kg}^{-1}\,\text{s}^{-2}.$$

In more complex expressions such as

$$\text{Temperature} = az^4 + bz^3 + cz^2 + dz + e \tag{2.8}$$

all terms must have the same units. Thus, since the left-hand side of this equation has units of °C, the units of az^4, bz^3, cz^2, dz and e must also be °C. Now, the units of z are kilometres and therefore the units of the constants must be as follows

$$\begin{aligned}\text{Units of } a &= {}^\circ\text{C km}^{-4}\\ \text{Units of } b &= {}^\circ\text{C km}^{-3}\\ \text{Units of } c &= {}^\circ\text{C km}^{-2}\\ \text{Units of } d &= {}^\circ\text{C km}^{-1}\\ \text{Units of } e &= {}^\circ\text{C}.\end{aligned}$$

In many textbooks you will see a slightly different way of doing dimensional analyses. In this approach, the **dimensions** of a unit are expressed in terms of mass, length and time (abbreviated to M, L and T). For example, acceleration has units of m s^{-2} which is a length divided by a time squared. Thus, acceleration has dimensions LT^{-2}. If this procedure is repeated for all items in equation (3.13), the result is

Dimensions of ρ

$$= \frac{\text{Dimensions of } 3 \times \text{Dimensions of } g}{\text{Dimensions of } 4 \times \text{Dimensions of } G \times \text{Dimensions of } \pi \times \text{Dimensions of } r}$$

$$= \frac{LT^{-2}}{L^3M^{-1}T^{-2}L} = ML^{-3} \qquad (4.43)$$

which balances since the dimensions of density are indeed mass divided by length cubed. In practice, it does not much matter which of these two approaches to dimensional analysis you use since they are entirely equivalent. However, I prefer the first in most situations since it is easier when quantities such as temperature are included and it also forces you to check that you have used consistent units for all quantities in the equation.

Special cases, the final checking method, is also useful for testing whether the general form for your answer is reasonable. Essentially, an expression is evaluated for a situation in which the correct answer is already known. In the radioactive dating method, for example, a specimen which formed very recently should still retain, more or less, the whole of its initial proportion of the radioactive element. Thus in equation (4.14)

$$Q = Q_0 \exp(-t/\lambda) \qquad (4.14)$$

we should obtain a value for Q equal to Q_0 when $t = 0$. Now, substituting $t = 0$ into this expression yields

$$\begin{aligned} Q &= Q_0 \exp(0) \\ &= Q_0 \times 1 \\ &= Q_0 \end{aligned}$$

as required. If, on the other hand, an equation of the form

$$Q = Q_0[1.0 - \exp(-t/\lambda)] \qquad (4.44)$$

had been suggested for the way that Q reduces with time, this yields a value of

$$Q = 0.0$$

in the case of $t = 0$. Since this is not the answer we expect, equation (4.44) must be wrong. N.B. As with dimensional analysis, it is possible for an incorrect answer to pass this test but a failure of the test means that the equation is definitely wrong (i.e. we can always say that an answer is definitely wrong but can never say that an answer is definitely right).

Question 4.6 Check that the equation for temperature, T, as a function of depth, z,

$$T = az + T_0$$

gives a sensible answer for the temperature at zero depth. In this expression, a is the temperature gradient and T_0 is the surface temperature.

FURTHER QUESTIONS

4.7 The thickness of a bottomset bed at the foot of a delta can often be well approximated by the expression

$$t = t_0 \exp(-x/X)$$

where t is thickness, x is distance from the bottomset bed start and t_0 and X are constants.

(i) Rearrange this expression into an equation for $\ln(t_0)$.
(ii) In a particular delta bottomset bed, the thicknesses are 5 and 0.1 m at $x = 1$ and 4 km respectively. Substitute these values into your answer from (i) to give two equations. Hence, obtain and solve an expression for X.
(iii) Estimate the bottomset thickness at $x = 0$ km.

4.8 The temperature, T, in the earth may be well approximated by an equation of the form

$$T = ar^4 + br^2 + c$$

where T is temperature, r is distance from the earth's centre and a, b and c are constants. At distances from the centre of 1260, 5660 and 6260 km the temperatures are 4300, 1900 and 1150 °C respectively. Use these figures to determine suitable values for a, b and c. Does the resultant expression give sensible values for temperature at the earth's centre and at the earth's surface?

4.9 If temperature, T (in °C) at a depth of z (in km) is approximately given by

$$T = az^2 + bz + c$$

determine the units of a, b and c.

4.10 Based upon dimensional analysis and special cases, which of the following equations, for sediment ages in a lake bed, are definitely wrong?

(i) Age = (Depth × Rate) + Age of top

(ii) $\text{Age} = \dfrac{\text{Rate}}{\text{Depth}} + \text{Age of top}$

(iii) $\text{Age} = \dfrac{\text{Depth}}{\text{Rate}} + \text{Depth of top}$

(iv) $\text{Age} = \dfrac{\text{Depth}}{\text{Rate}} + \text{Age of bottom bed}$

where all ages are in ky, rate is in m ky^{-1} and all depths are in metres.

Trigonometry

5

5.1 INTRODUCTION

Trigonometry is the study of triangles. Triangles rather than, say, squares or hexagons because any other **polygon** (a closed shape with straight edges) can be constructed by adding triangles together (Figure 5.1). Thus, if the properties of triangles are understood, any other polygon can also be dealt with.

Triangles are ideal for purposes such as mapping since there is a simple set of rules relating the lengths of their sides to the size of their angles. Figure 5.2 illustrates the quantities which define a given triangle. This triangle has three sides of length a, b and c and three angles of size α, β and γ. Letters A, B and C are used here to denote the **vertices** (i.e. corners) of the triangle. This gives an alternate way of specifying angles. For example, angle α could be denoted angle BAC, i.e. the angle formed between line BA and line AC.

Question 5.1 Using a ruler and protractor, sketch the following triangles and determine the unknown three quantities:

(a) $\alpha = 20°, \gamma = 100°, a = 4\,\text{cm}$.

(b) $\gamma = 20°$, $a = 3\,\text{cm}$, $b = 5\,\text{cm}$.

Question 5.2 Examine the map in Figure 5.3 and measure the distance from the church to the transmitter. If, from an exposure, the church is seen to be located 45° W of N whilst the transmitter is due west, where is the exposure? How far is the exposure from the church and how far from the transmitter?

Angles, in geology, are normally measured in degrees since this is a convenient unit for measuring dips, strikes and other similar quantities. However, there are other units which can be used of which **radians** are the most important. An angle of one radian is about 57.3°. This may seem a very peculiar, and rather large, unit but there are good reasons for its use one of which will be explained in Chapter 8. For now, I will only point out that the radian is defined

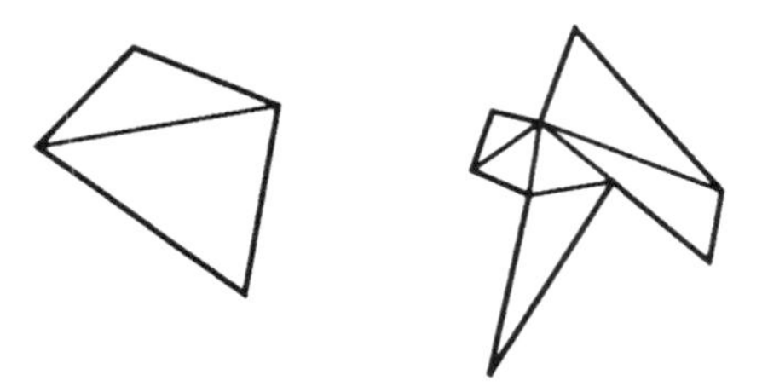

Figure 5.1 Any polygon can be constructed from a set of triangles.

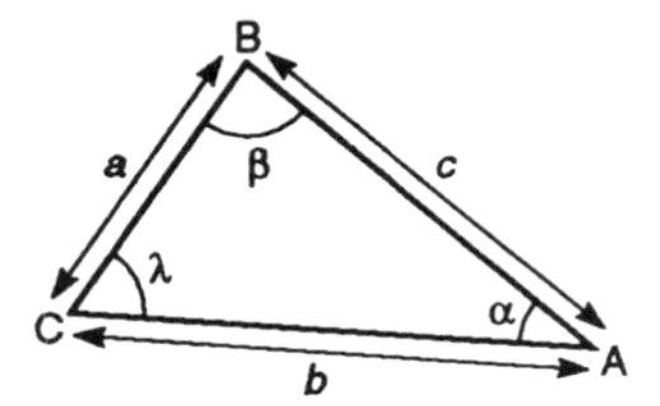

Figure 5.2 A triangle is described by the lengths of the three sides (a, b and c here) and the size of the three angles (α, β and γ). The vertices of the triangle are denoted A, B and C.

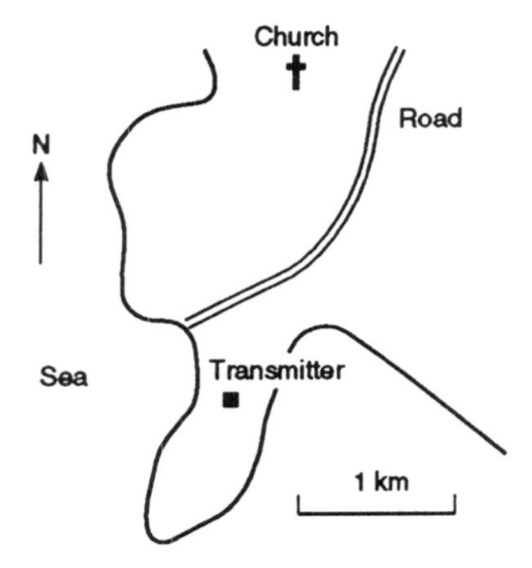

Figure 5.3 Locate the exposure given the information in question 5.2.

such that one complete rotation (i.e. an angle of 360°) is 2π radians (~ 6.28 radians).

Question 5.3 Given that 360° is equivalent to 2π radians, what are the following angles in radians? (Hint: what fraction of a complete rotation are these angles?)

(i) 180°, (ii) 90°, (iii) 270°, (iv) 100°.

5.2 TRIGONOMETRIC FUNCTIONS

Throughout this chapter you will be using the sine, cosine and tangent functions. These are collectively known as **trigonometric functions.** They are usually abbreviated in equations and tables to sin, cos and tan respectively. What are these functions and why are they useful?

Figure 5.4 shows two **right-angled triangles** (triangles in which one angle is 90°) each of which contains the same angle, θ. However, the second triangle has sides which are k times longer than those of the first, i.e. $o_2 = ko_1$, $a_2 = ka_1$ and $h_2 = kh_1$ where k is a constant. Triangles such as these, which are exactly the same shape but which are of different sizes, are known as *similar* triangles. Incidentally, I have denoted the lengths of the sides using o because this is the side *o*pposite the given angle, a because this is the side *a*djacent to θ and h for ***h*ypotenuse** which is the side opposite the right angle.

Now, for the larger triangle, the length of the opposite side divided by the length of the adjacent side is

$$\begin{aligned} o_2/a_2 &= (ko_1)/(ka_1) \\ &= o_1/a_1 \end{aligned} \tag{5.1}$$

i.e. dividing the length of the opposite side by the length of the adjacent side gives the same value for both triangles. This value will only depend upon the angle θ. This ratio is called the *tangent* of θ (or $\tan(\theta)$) and can be found either by looking it up in tables or by the use of a calculator. Thus

$$\tan(\theta) = \text{Length of the opposite side/Length of the adjacent side.} \tag{5.2}$$

It is worth having a look at the ratios formed from pairs of sides other than a and o in Figure 5.4. For example, the length of the opposite side divided by

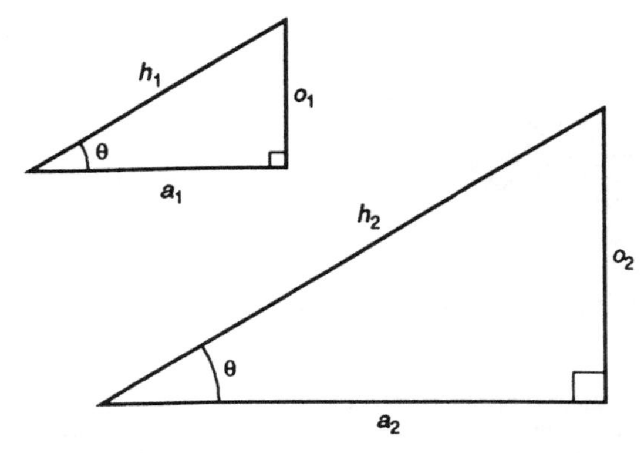

Figure 5.4 Two similar triangles. The lower triangle has sides which are k times longer than those of the upper triangle. However, all angles are identical.

the length of the hypotenuse is also the same for both triangles since.

$$o_2/h_2 = (ko_1)/(kh_1)$$
$$= o_1/h_1. \quad (5.3)$$

This ratio, again, only depends upon the angle θ and is called the sine of θ or

$$\sin(\theta) = \text{Length of the opposite side/Length of the hypotenuse.} \quad (5.4)$$

Finally, the length of the adjacent side divided by the length of the hypotenuse is a constant for the two triangles since

$$a_2/h_2 = (ka_1)/(kh_1)$$
$$= a_1/h_1. \quad (5.5)$$

This ratio is called the cosine of θ, i.e.

$$\cos(\theta) = \text{Length of the adjacent side/Length of the hypotenuse.} \quad (5.6)$$

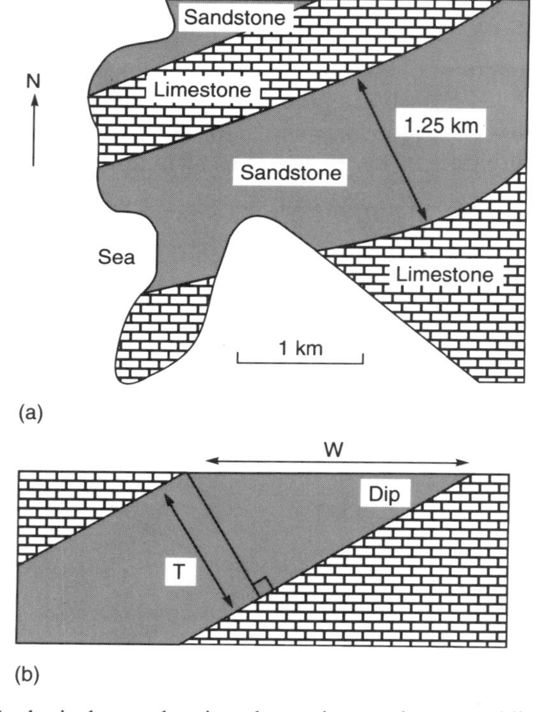

Figure 5.5 (a) Geological map showing alternating sandstone and limestone bedding. One of the sandstone formations has a width of 1.25 km and a dip of 27°. (b) Vertical cross-section through a dipping bed which has a true thickness T and an apparent thickness projected on the surface of W.

N.B. The definitions of tan, cos and sine given above are only true for right-angled triangles.

Question 5.4 The hypotenuse of a right-angled triangle is twice the length of one of the other sides. Calculate cos, sine and tan for one (non-right-angled) angle in the triangle. (Hint: let one side have a length x giving a hypotenuse of length $2x$. Then use Pythagoras' theorem (i.e. $h^2 = a^2 + o^2$) to find the length of the third side. You will probably find a sketch helpful as well.)

What are these functions used for? Figure 5.5 illustrates a common situation in which the sine function can be used. The geological map (Figure 5.5(a)) shows an alternating sequence of sandstone and limestone formations. One of the sandstone formations has an outcrop width of 1.25 km and its beds dip at 27°. What is the true thickness of this formation? Figure 5.5(b) shows how the apparent width, W, of a bed or formation is related to its true thickness, T, and its dip. From the definition of sine (equation 5.4) it follows that

$$\sin(\text{Dip}) = T/W \tag{5.7}$$

which, after rearrangement, yields

$$T = W\sin(\text{Dip}). \tag{5.8}$$

Substituting the known values for W and the dip and using a calculator (or tables) to calculate $\sin(\text{Dip})$, gives

$$\begin{aligned} T &= 1.25\sin(27°) \\ &= 1.25 \times 0.454 \\ &= 0.567\,\text{km} \\ &= 567\,\text{m}. \end{aligned} \tag{5.9}$$

Question 5.5 A cliff has a height of 130 m. A particular sedimentary bed outcrops at the cliff top and dips at 42.5° in a direction parallel to the cliff edge. Draw a sketch of this and, by considering the definition of the tangent function, determine how far away, horizontally, the same bed outcrops at the cliff base.

The **inverse trigonometric functions** produce the angle corresponding to a particular value for a sine, cosine or tangent. For example, $\sin(37°) = 0.602$ and therefore the inverse sine of 0.602 equals 37°. The inverse tangent, sine and cosine functions are sometimes called the arctangent, arcsine and arcosine functions. In equations they are denoted by $\tan^{-1}$, $\sin^{-1}$ and $\cos^{-1}$ respectively (e.g. $\sin^{-1}(0.602) = 37°$). Some calculators use a notation of atan, asin and acos instead or, very occasionally, arctan, arcsin and arcos.

The standard notation is very poor since there is a very similar notation for denoting powers of trigonometric functions. For example, the square of $\tan(\theta)$ (i.e. $\tan(\theta)\tan(\theta)$) is usually written $\tan^2(\theta)$. Thus, $\tan^{-1}(\theta)$ might be thought, erroneously, to be the same as $1/\tan(\theta)$. Unfortunately, this way of denoting the inverse trigonometric functions is very well established and is unlikely to be dropped now.

The fact that, with these inverse functions, angles can now be found from knowledge of their sines, cosines or tangents greatly increases the power of trigonometry. For example, the inverse tangent function can be used to determine true bed dips from a cross-section which has vertical exaggeration. Geological cross-sections frequently have different scales in the vertical and horizontal directions since data may be mapped over several kilometres horizontally but only extrapolated downwards for a few hundred metres. Figure 5.6 shows an example in which the cross-section is 4 or 5 km wide but only about 100 m deep. The vertical exaggeration here is about 12 to 1 (i.e. the vertical scale is stretched 12-fold relative to the horizontal scale). Thus, to get a section in true scale, all vertical distances should be shrunk by a factor of 12. As a result, the beds, which appear to have a dip of about 30°, have a true dip which is much less. The true dip may be found by noting that, from equation (5.2) and Figure 5.6,

$$\begin{aligned}\tan(\text{Dip}) &= \text{Opposite}/\text{Adjacent}\\ &= 100\,\text{m}/2\,\text{km} = 100/2000\\ &= 0.05\end{aligned} \tag{5.10}$$

therefore, using the inverse tangent,

$$\begin{aligned}\text{Dip} &= \tan^{-1}(0.05)\\ &= 2.86^\circ.\end{aligned} \tag{5.11}$$

Thus, the true dip is less than 3°, i.e. about one-tenth of the apparent dip in Figure 5.6.

> Question 5.6 A cliff has a height of 45 m. A bed at the cliff top outcrops 110 m, horizontally, from where it outcrops at the cliff base. Draw a sketch and determine the value of the tangent of the bed dip. Using the inverse tangent, find the dip in degrees.

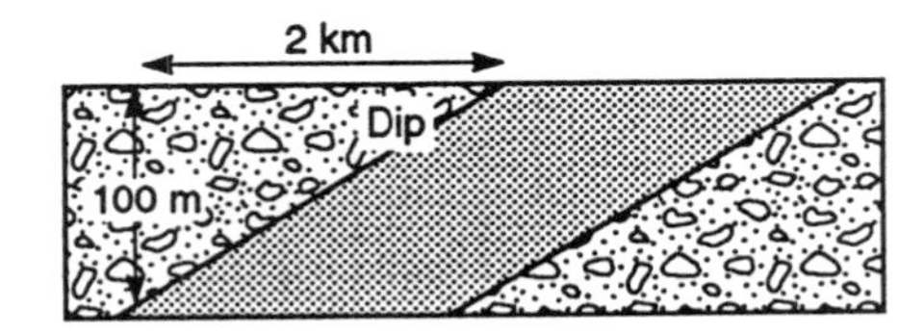

Figure 5.6 Simplified geological cross-section in which the vertical scale is around 12 times larger than the horizontal scale. The result is an apparent bed dip which is much greater than the true dip.

5.3 DETERMINING UNKNOWN ANGLES AND DISTANCES

In questions 5.1 and 5.2 earlier in this chapter, the angles and side lengths of several triangles were determined by drawing a sketch using the supplied information and measuring the unknown lengths and angles. Clearly, it would be more convenient and more accurate if the unknowns could be calculated, rather than measured, and this is indeed possible.

To find unknown values in a triangle, three rules are needed:

1. *The 180° rule.* The angles must add up to exactly 180°. This rule allows us to find the third angle whenever two of the angles are known.
2. *The sine rule.* For a given triangle, the length of any side divided by the sine of the opposite angle is a constant. In terms of the symbols defined in Figure 5.2 this becomes

$$\frac{a}{\sin \alpha} = \frac{b}{\sin \beta} = \frac{c}{\sin \gamma}. \quad (5.12)$$

3. *The cosine rule.* This is a generalization of Pythagoras' theorem to cover non-right-angled triangles. In terms of the symbols defined in Figure 5.2, the cosine rule is

$$a^2 = b^2 + c^2 - 2bc\cos(\alpha) \quad (5.13)$$

or

$$b^2 = a^2 + c^2 - 2ac\cos(\beta) \quad (5.14)$$

or

$$c^2 = a^2 + b^2 - 2ab\cos(\gamma). \quad (5.15)$$

Question 5.7 Using the symbols from Figure 5.2:

(a) If $\alpha = \beta = 70°$, use the 180° rule to find angle γ.
(b) If $b = 3$ km, $c = 2$ km and $\gamma = 40°$, use the sine rule to find angle β.
(c) If $b = 3$ km, $c = 1$ km and $\alpha = 37°$, use the cosine rule to find length a.

Question 5.8 If angle α is a right angle show that equation (5.13) reduces to Pythagoras' theorem $a^2 = b^2 + c^2$.

The 180° rule, the sine rule and the cosine rule are used in different ways and different orders depending upon the information known at the start of a specific problem. Figure 5.7 illustrates the four possible starting points. Our starting point could be knowledge of: three sides (Figure 5.7(i)); two sides plus one angle (Figures 5.7(ii) and (iii)); or one side and two angles (Figure 5.7(iv)). There are two cases to consider when two sides plus one angle are known since the angle might be that formed between the two known sides (Figure 5.7(ii)) or might be opposite one of the known sides (Figure 5.7(iii)). These two cases must be treated differently and, in fact, the case of two known sides plus an opposite angle has two solutions (this will be explained later).

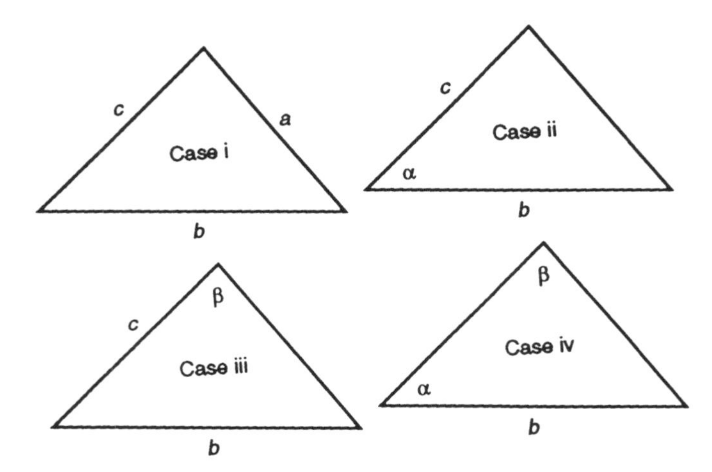

Figure 5.7 The four cases to consider when finding unknown quantities in triangles: (i) three sides and no angles known; (ii) two sides and one included angle known; (iii) two sides and one opposite angle known; (iv) one side and any two angles known.

Figure 5.8 is a flow chart which shows a scheme for finding all the unknown quantities in a triangle given the four starting points shown in Figure 5.7. Other schemes are possible and may be more suitable in some cases but this scheme should always work.

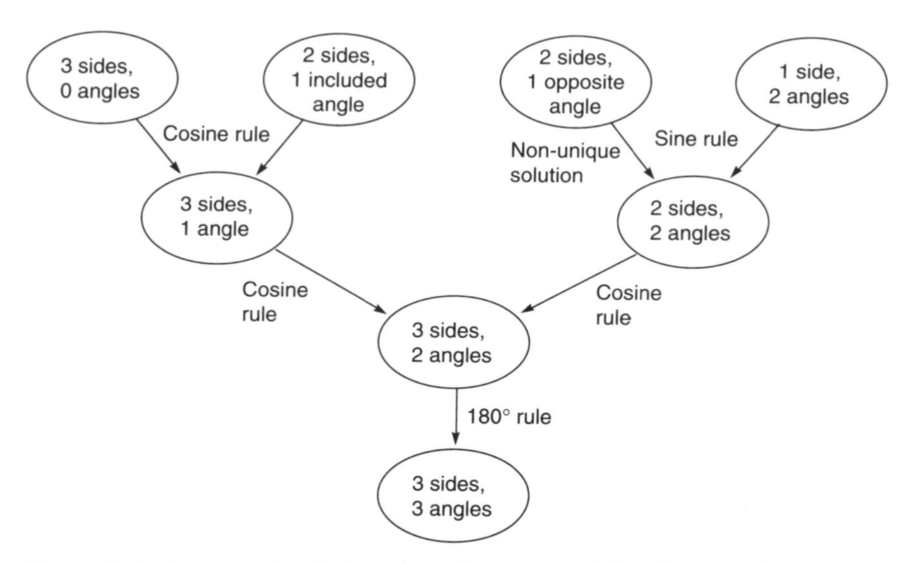

Figure 5.8 A flow chart for finding the unknown quantities given the four starting points shown in Figure 5.7. Note that starting with two sides and an opposite angle gives two possible solutions.

Suppose that three sides and no angles are known. From the flow chart, the first step is to use the cosine rule to find one of the angles. Equation (5.13) can be rearranged into

$$\cos(\alpha) = \frac{b^2 + c^2 - a^2}{2bc} \tag{5.16}$$

which can then be used to find angle α by using the inverse cosine function to give

$$\alpha = \cos^{-1}[(b^2 + c^2 - a^2)/2bc]. \tag{5.17}$$

Three sides and one angle, α, are now known and the flow chart suggests that the next step should be to use the cosine rule again to find a second angle. Rearranging equation (5.14) gives

$$\cos(\beta) = \frac{a^2 + c^2 - b^2}{2ac} \tag{5.18}$$

and then, using the inverse cosine function

$$\beta = \cos^{-1}[(a^2 + c^2 - b^2)/2ac]. \tag{5.19}$$

The final unknown quantity is the angle γ which, using the 180° rule, is found from

$$\gamma = 180 - \alpha - \beta. \tag{5.20}$$

Question 5.9 Two exposures, 500 m apart, are 400 and 200 m, respectively, from a church. Calculate, using equations (5.17), (5.19) and (5.20), the angles contained by the triangle defined by the two exposures and the church.

Question 5.10 Using the flow chart (Figure 5.8), find the unknown quantities in the following:

(i) $\alpha = 40°, b = 5$ km, $c = 2$ km.
(ii) $\alpha = 40°, b = 3$ km, $a = 2$ km.
(iii) $\alpha = 40°, \beta = 60°, a = 3$ km.

5.4 CARTESIAN COORDINATES AND TRIGONOMETRIC FUNCTIONS OF ANGLES BIGGER THAN 90°

The definitions of the trigonometric functions given in section 5.2 are only valid for angles less than 90°. However, triangles frequently have one angle greater than this. So, how are these functions defined in these cases? It is easiest to begin by first discussing Cartesian coordinates (Figure 5.9). This is a way of specifying any location in a plane by giving the horizontal and vertical distance from an **origin** (the centre of the coordinate system where $x = y = 0$).

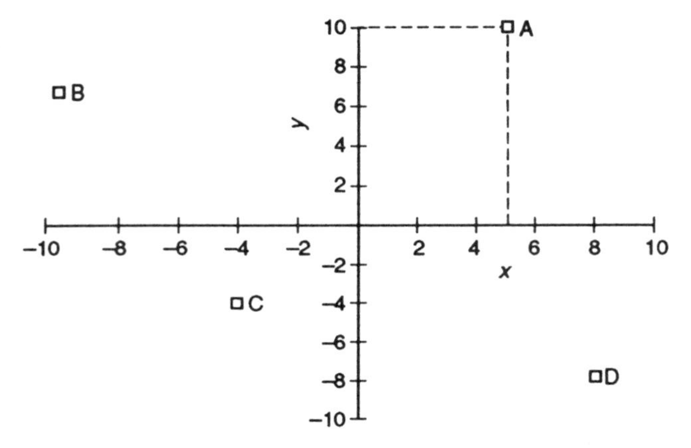

Figure 5.9 Cartesian coordinates to specify the locations of points. Point A is at position $x = 5$, $y = 10$; point B is at $x = -10$, $y = 7$; point C at $x = -4$, $y = -4$; point D at $x = 8$, $y = -8$.

Point A, for example, is at the location $x = 5$, $y = 10$. This is frequently abbreviated to 'the point (5,10)'. Note that points to the left of the origin have a negative x-coordinate and points below the origin have a negative y-coordinate.

This same coordinate system could be used to specify angles by drawing lines between the origin and these points (Figure 5.10). The corresponding angle is that formed between the x-axis and the line, measured in an anticlockwise sense. (N.B. When measuring compass bearings, angles are

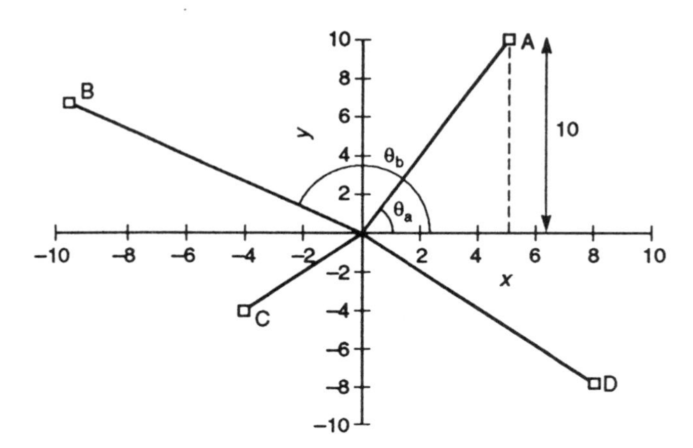

Figure 5.10 Using Cartesian coordinates to specify angles. The angles of interest are formed between the x-axis and lines joining the specified points to the origin. All angles are measured anticlockwise. To keep the diagram uncluttered, only two angles are shown here.

measured clockwise around from north. It is unfortunate but mathematicians and cartographers have settled on opposite conventions and you will have to get used to using these different conventions in different contexts.)

Starting with the point A: the angle, θ_a, corresponding to the point (5,10), has a tangent of

$$\begin{aligned}\tan(\theta_a) &= \text{Opposite/Adjacent}\\ &= y\text{-coordinate}/x\text{-coordinate}\\ &= 10/5\\ &= 2.0\end{aligned} \qquad (5.21)$$

i.e. the angle is $\tan^{-1}(2) = 63.4°$. Now, the same procedure could be attempted with point B at (−10,7). The angle of interest is now between 90° and 180°. Note that x is therefore negative. In other words, the length of the adjacent side is negative giving

$$\begin{aligned}\tan(\theta_b) &= \text{Opposite/Adjacent}\\ &= 7/-10\\ &= -0.7.\end{aligned} \qquad (5.22)$$

thus for this case the tangent is a negative number.

Question 5.11 Repeat the above procedure to find the tangents of the angles produced using points C and D in Figure 5.10.

The same process could now be attempted for the cosines and sines of the angles produced by points A, B, C and D. For the sine and cosine calculations the hypotenuse is the line from the origin to the point and is always taken to have a positive length. Figure 5.11 summarizes the results by displaying sine, cosine and tangent for angles between 0° and 360°.

Note that there is more than one angle which gives rise to any particular value for the sine, cosine or tangent (e.g. $\tan(45°) = \tan(225°) = 1.0$). For this reason, the angles obtained from calculating the inverse trigonometric functions are not unique. Thus, your calculator would give $\tan^{-1}(1.0) = 45°$ but the answer could be 225°. In general, there are two angles between 0° and 360° which give rise to any given value for sine, cosine or tangent. You should be aware that this non-uniqueness can, occasionally, result in incorrect answers. Consider Figure 5.12 in which a triangle is shown with an unknown angle, θ, clearly much larger than 90°.

The obvious way to determine θ is to use the sine rule which leads to

$$\sin(20)/5 = \sin(\theta)/10 \qquad (5.23)$$

thus

$$\begin{aligned}\sin(\theta) &= 2\sin(20)\\ &= 0.684\end{aligned} \qquad (5.24)$$

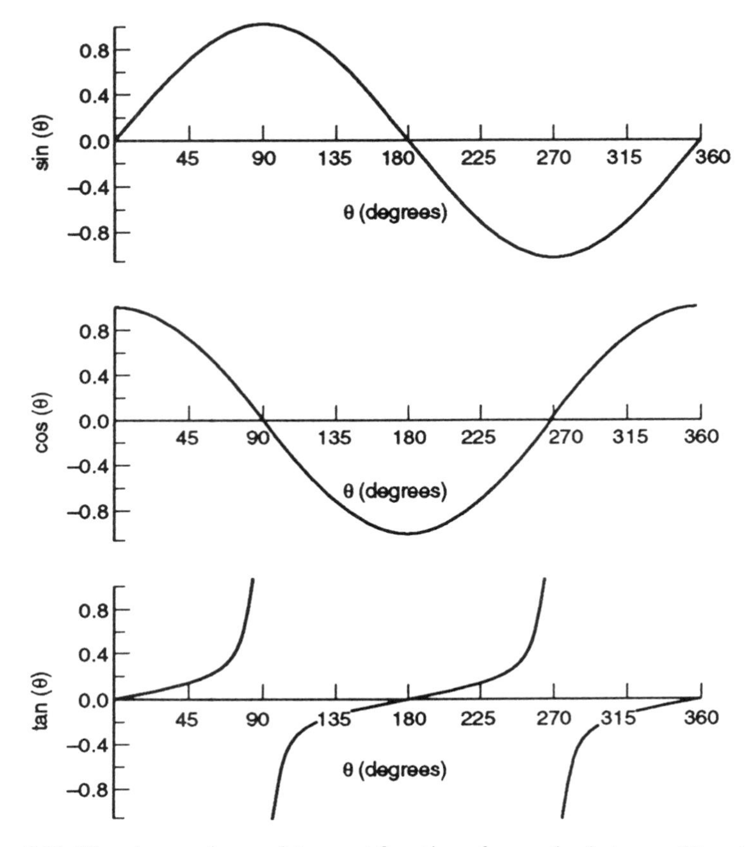

Figure 5.11 The sine, cosine and tangent functions for angles between 0° and 360°.

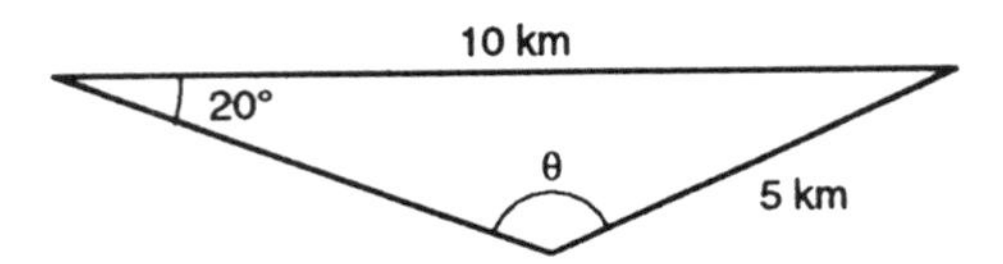

Figure 5.12 What size is θ in this triangle?

giving

$$\begin{aligned} \theta &= \sin^{-1}(0.684) \\ &= 43.2^\circ \end{aligned} \tag{5.25}$$

which is plainly wrong. The reason for this is that there are two angles between 0° and 180° which have a sine of 0.684. Inspecting Figure 5.13, an angle of $180 - 43.2$ degrees has the same sine as 43.2°. Thus, the correct answer is 136.8°.

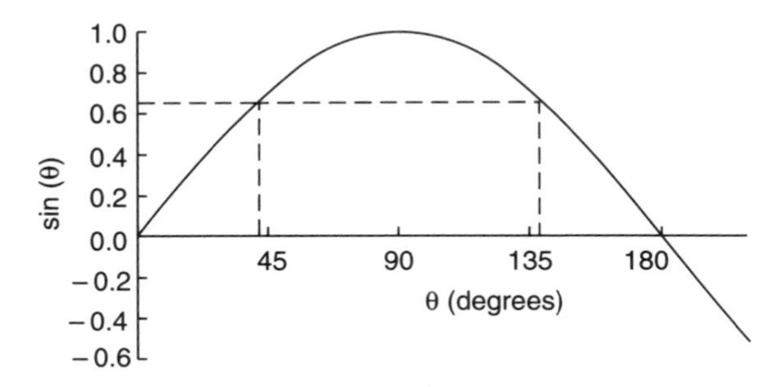

Figure 5.13 Non-uniqueness of the inverse trigonometric functions means that, for example, $\sin^{-1}(0.684)$ could be 43.2° or it could be 136.8° (i.e. 180 − 43.2).

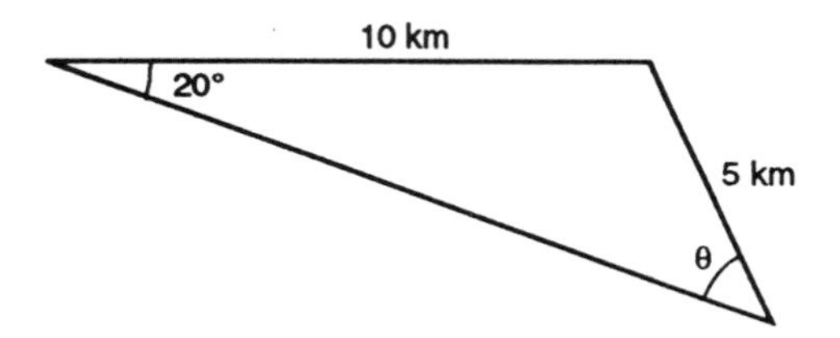

Figure 5.14 Compare this to Figure 5.12. It has exactly the same known data (two sides and one opposite angle) but the solution to θ is very different.

The reason for this uncertainty is simply that the other answer, 43.2°, is possible with the information given. This is illustrated by Figure 5.14 which has exactly the same known starting information (two sides and one opposite angle) but which actually has a solution of 43.2°. In the case of two sides and one opposite angle, we must also know if the other opposite angle is less than or greater than 90°.

In conclusion, whenever you use the inverse sine, cosine or tangent functions, calculate both solutions and then decide which is appropriate in the particular case you are investigating.

5.5 TRIGONOMETRY IN A THREE-DIMENSIONAL WORLD

Up to now, I have used trigonometry purely in two dimensions. However, geology is a three-dimensional subject. Figure 5.15(a) illustrates a typical three-dimensional problem. Imagine a dipping bedding plane outcropping on a cliff face which is not parallel to the direction of maximum slope. This will result in an apparent dip, on the cliff face, which is less than the true dip. The most extreme case is where the cliff face is at right angles to the direction of dip (i.e. the cliff is in the **strike** direction). In this extreme case the apparent

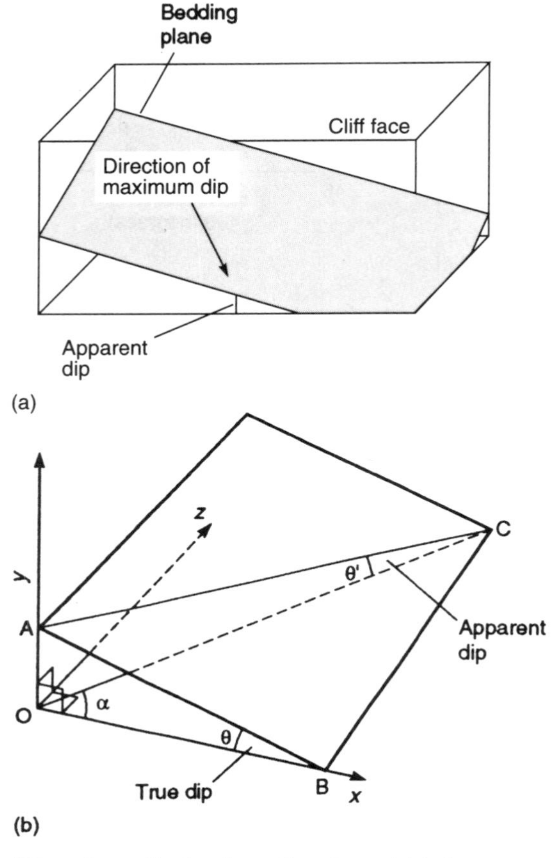

Figure 5.15 (a) A bedding plane dipping in a direction which is not parallel to the cliff face which cuts through it. This results in an apparent dip on the cliff face which is less than the true dip. (b) Construction for determining the relationship between true dip and apparent dip. The x–y plane in (b) is parallel to the direction of maximum dip in (a). Triangle OAC is then parallel to the cliff face.

dip of the beds is zero! Is there a simple relationship between the true dip and the apparent dip?

Figure 5.15(b) shows a construction for determining this relationship. In this diagram the bedding plane has a true dip, θ, in a direction parallel to the x-axis. The line OC represents the direction along which the bedding plane is cut (i.e. the cliff face). This direction is at an angle α to the dip direction. This results in an apparent dip of θ'. Note that the angles COA, BOA and OBC are all right-angles. From this it follows that

$$\tan(\theta) = \mathrm{OA/OB} \tag{5.26}$$

$$\tan(\theta') = \mathrm{OA/OC} \tag{5.27}$$

and

$$\cos(\alpha) = \text{OB}/\text{OC}. \quad (5.28)$$

Equations (5.27) and (5.28) can be combined as follows:

$$\begin{aligned}\tan(\theta') &= \text{OA}/\text{OC} = (\text{OA}/\text{OB})(\text{OB}/\text{OC}) \\ &= (\text{OA}/\text{OB})\cos(\alpha) \quad (5.29)\end{aligned}$$

which, together with equation (5.26) leads to

$$\tan(\theta') = \tan(\theta)\cos(\alpha) \quad (5.30)$$

which relates the apparent dip, θ', to the true dip, θ, and the angle, α, which the cliff makes to the dip direction. From this, the apparent dip can be found using

$$\theta' = \tan^{-1}[\tan(\theta)\cos(\alpha)] \quad (5.31)$$

or the true dip can be calculated from the apparent dip using

$$\theta = \tan^{-1}[\tan(\theta')/\cos(\alpha)]. \quad (5.32)$$

If an apparent dip is 32° measured in a direction 25° from the direction of maximum dip, the true dip is

$$\begin{aligned}\theta &= \tan^{-1}[\tan(32)/\cos(25)] \\ &= \tan^{-1}[0.625/0.906] \\ &= \tan^{-1}[0.689] \\ &= 34.6°. \quad (5.33)\end{aligned}$$

Question 5.12 On a cliff face, the apparent dip is 25° whilst the true dip is 35°. What is the angle between the cliff face and the strike direction?

Finding the true dip and its direction, in the field, can become even more complex than indicated. Additional difficulties not considered so far are such things as the effect of uneven topography, non-planar bedding and measurements made on inclined surfaces. Fortunately, there is a much easier approach using **stereographic projection**, a subject that will be introduced in Chapter 6.

5.6 INTRODUCTION TO VECTORS

This section will give a brief introduction to **vectors**. A more detailed treatment is beyond the scope of this book but the most common vector operation (**vector addition**) is covered.

A vector is any quantity which has a direction as well as a magnitude. The flow of water in a river channel, for example, can be described in terms of its direction and rate of flow. Another example is the earth's magnetic field which, at any given point on the earth's surface, has a definite direction (roughly speaking the field points north with a dip which depends upon latitude) as well as a definite strength (the strength increases towards the poles). Other quantities which have only a magnitude but no direction are called **scalar quantities** (e.g. temperature).

> Question 5.13 Are the following vector or scalar quantities?
> (i) mass, (ii) gravitational acceleration, (iii) shear strength, (iv) age, (v) the line joining an exposure location to a church.

Figure 5.16 shows a series of vectors representing flow at various locations on

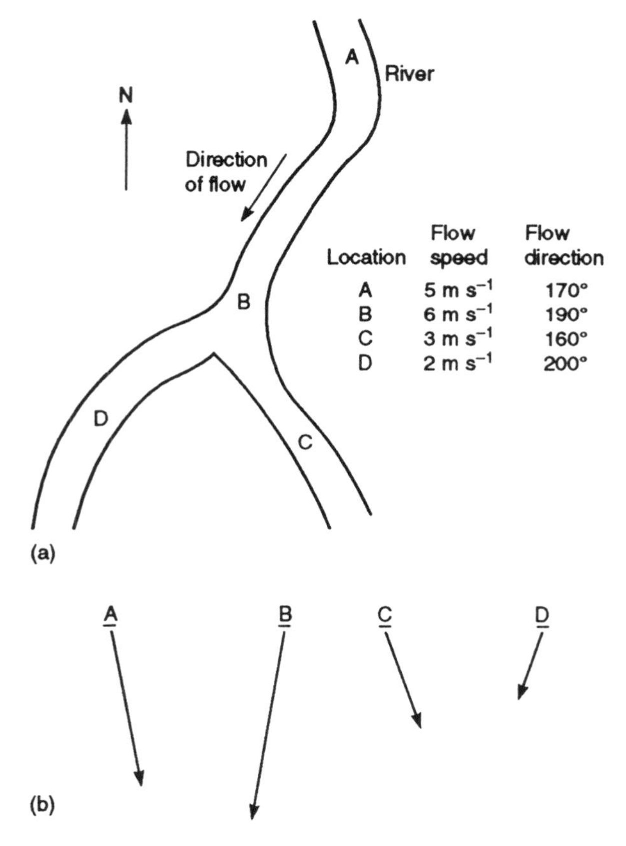

Figure 5.16 (a) A river flowing, roughly, southwards. At points A, B, C and D the river speed and direction are as shown in the table. (b) Vector representation of the river flow at points A, B, C and D. The direction of the arrows is in the direction of flow and the length of the arrow is proportional to the flow speed. Thus, vector $\underline{B}$ is the longest and vector $\underline{D}$ is the shortest.

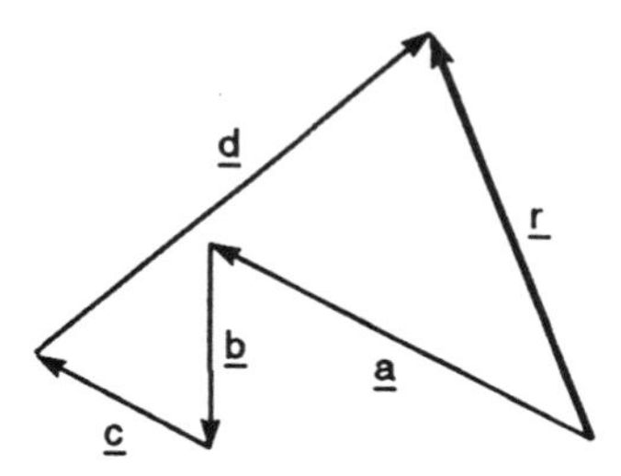

Figure 5.17 Vector addition. Vectors $\underline{a}$, $\underline{b}$, $\underline{c}$ and $\underline{d}$ are added together by placing them nose-to-tail as shown. The resultant vector is $\underline{r}$. Algebraically this is written $\underline{r} = \underline{a} + \underline{b} + \underline{c} + \underline{d}$.

a river. The arrows, representing the flow, point in the direction of flow and have a length which is proportional to the flow speed. These arrows are diagrammatic representations of the flow vectors. These vectors can be denoted by the letters $\underline{A}$, $\underline{B}$, $\underline{C}$ and $\underline{D}$ where underlining is a way of indicating that they are vectors. An alternative notation is to indicate vectors by using bold face (i.e. **A**, **B**, **C** and **D**). In this section, I shall deliberately alternate these so that you get used to seeing vectors written both ways. Obviously, if you are writing vector expressions by hand, it is easier to use the underlining convention.

An important property of vectors is that they may be added together. Vector addition is simply the process of combining the vectors 'nose-to-tail' as shown in Figure 5.17. The resultant vector is obtained by drawing a vector from the tail of the first vector to the nose of the last. This operation can be represented algebraically by the equation

$$\underline{r} = \underline{a} + \underline{b} + \underline{c} + \underline{d}. \tag{5.34}$$

An important point about this addition is that it is not the same as adding the vector lengths and vector directions separately.

Question 5.14 Draw a set of x–y axes on a sheet of paper. Then draw the vectors:

(i) Vector $\underline{a}$: length 3 cm, direction 10° clockwise from the x-axis,
(ii) Vector $\underline{b}$: length 5 cm, direction 50° clockwise from the x-axis,
(iii) Vector $\underline{c}$: length 3 cm, direction 190° clockwise from the x-axis,
(iv) The vector $\underline{d} = \underline{a} + \underline{b}$,
(v) The vector $\underline{e} = \underline{a} + \underline{c}$.

Another useful operation on vectors is **scalar multiplication** (do not confuse this with the **scalar product**, a more sophisticated vector operation, which will not be discussed further here). In scalar multiplication the vector length is

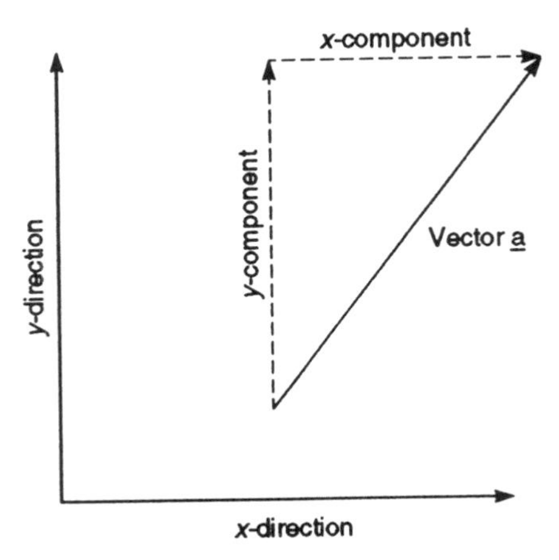

Figure 5.18 A vector can be thought of as the sum of two component vectors. One of these is in the x-direction and is called the x-component. The other is in the y-direction and is called the y-component.

simply increased by multiplying it by a scalar. Thus, a vector in the direction 13° E of N with a length (or **magnitude**) of 5 km becomes, after scalar multiplication by 3, a vector in the direction 13° E of N with a magnitude of 15 km. A small complication is the effect of multiplying by a negative quantity (e.g. -3). In this case, the direction of the vector is reversed and so a vector in the direction 13° E of N with a magnitude of 5 km becomes, after scalar multiplication by -1, a vector in the direction 193° E of N with a magnitude of 5 km. Similarly, after multiplying by -3, a vector in the direction 13° E of N with a magnitude of 5 km becomes a vector in the direction 193° E of N with a magnitude of 15 km.

Vector addition and scalar multiplication together allow a new way of specifying a vector to be introduced. The idea is to specify the vector in terms of the lengths of **component vectors** in the x- and y-directons. This is illustrated in Figure 5.18. The sum of the x-component and the y-component vectors produce the given vector. The vector can thus be written down as

$$\mathbf{a} = x\mathbf{i} + y\mathbf{j} \tag{5.35}$$

where $\mathbf{i}$ and $\mathbf{j}$ are vectors of length 1.0 in the x- and y-directions respectively and where x and y are the lengths of the x- and y-component vectors. Thus, $x\mathbf{i}$ is a vector of length x in the x-direction and $y\mathbf{j}$ is a vector of length y in the y-direction. In other words, vector $\mathbf{a}$ is the sum of the x- and y-component vectors. Vectors such as $\mathbf{i}$ and $\mathbf{j}$, which are of unit length, are known as **unit vectors**.

To change a vector specified in terms of magnitude and direction into a vector specified in terms of its x- and y-components, it is only necessary to use a little trigonometry. From Figure 5.18, the lengths of the x- and y-components are

$$x = a\cos(\theta) \tag{5.36}$$

and

$$y = a\sin(\theta) \tag{5.37}$$

where a is the length of vector **a** and θ is the angle which vector **a** makes to the x-direction. Thus if the length, a, and direction, θ, of the vector are known, the x- and y-components can be easily found. To convert from components to vector magnitude, Pythagoras' theorem gives

$$\text{Vector length, } a = \sqrt{(x^2 + y^2)}. \tag{5.38}$$

The vector direction follows from the definition of the tangent function and, for the case of Figure 5.18, gives

$$\text{Vector direction, } \theta = \tan^{-1}(y/x). \tag{5.39}$$

Question 5.15 Use the definitions of the sine and cosine functions and Figure 5.18 to derive equations (5.36) and (5.37) above.

The advantage of recasting vectors in terms of their components is that it makes addition much simpler. To add vectors together, you simply add the components. An example is illustrated in Figure 5.19. In this example there are two closely spaced faults with different throws and different fault dips. The question is: what is the total movement of block 1 relative to block 3? Using vectors makes this problem extremely straightforward. All we do is add the slip vector (i.e. a vector representing the direction of slip and amount of throw) for fault 1 to the slip vector for fault 2. So, what are the slip vectors, $\mathbf{s}_1$ and $\mathbf{s}_2$, for each of the faults? Using trigonometry in an identical manner to that

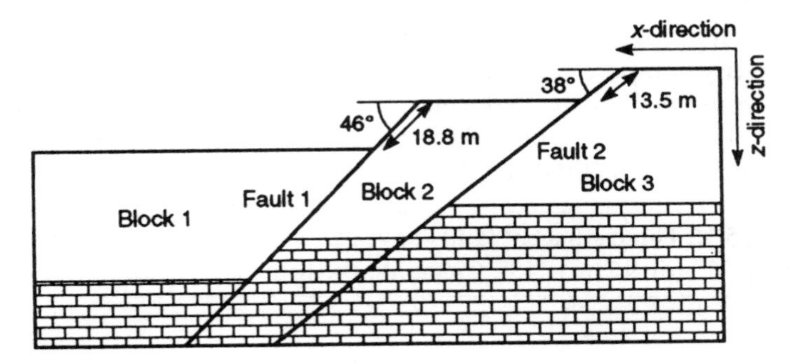

Figure 5.19 Slip vectors for two faults. The throws and slip directions are different on the two faults shown. What is the overall slip of the lower block relative to the upper?

used for determining equations (5.36) and (5.37) leads to

$$\begin{aligned}\mathbf{s}_1 &= 18.8\cos(46^\circ)\mathbf{i} + 18.8\sin(46^\circ)\mathbf{k}\\ &= 13.1\mathbf{i} + 13.5\mathbf{k}\end{aligned} \tag{5.40}$$

and

$$\begin{aligned}\mathbf{s}_2 &= 13.5\cos(38^\circ)\mathbf{i} + 13.5\sin(38^\circ)\mathbf{k}\\ &= 10.6\mathbf{i} + 8.3\mathbf{k}\end{aligned} \tag{5.41}$$

where **k** is a unit vector in the z-direction. Summing these vectors gives a resultant, total slip, of

$$\begin{aligned}\mathbf{s} &= \mathbf{s}_1 + \mathbf{s}_2\\ &= (13.1 + 10.6)\mathbf{i} + (13.5 + 8.3)\mathbf{k}\\ &= 23.7\mathbf{i} + 21.8\mathbf{k}.\end{aligned} \tag{5.42}$$

However, we would probably wish to have the final answer in the form of the dip and throw of a single fault which would give the same effect. Thus, we need to convert back from vector components to vector direction and magnitude. Applying the same principles as used in equations (5.38) and (5.39) to the fault throw problem gives

$$\begin{aligned}\text{Total throw} &= \sqrt{(23.7^2 + 21.8^2)}\\ &= 32.2\,\text{m}\end{aligned} \tag{5.43}$$

and

$$\begin{aligned}\text{Equivalent single fault dip} &= \tan^{-1}(21.8/23.7)\\ &= 42.6^\circ.\end{aligned} \tag{5.44}$$

Thus, a single fault dipping at 42.6° with a throw of 32.2 m, would have given an identical vertical and horizontal movement to block 1 relative to block 3. The resultant dip direction is called a **vector mean** direction. This kind of analysis would be useful when examining the variation in extension across fault systems as we move along strike. The number, dip and throw of faults generally vary along strike but by summing the individual slip vectors we could directly compare the size and direction of extension. In general, this analysis would need to be performed as a problem in three dimensions in which case the slip vectors have three components rather than two. However, apart from this, the methods used would be identical.

Question 5.16 Three adjacent faults have throws and dips of: (i) 10 m at 60°; (ii) 5 m at 65°; (iii) 12 m at 45°. Calculate the total slip vector.

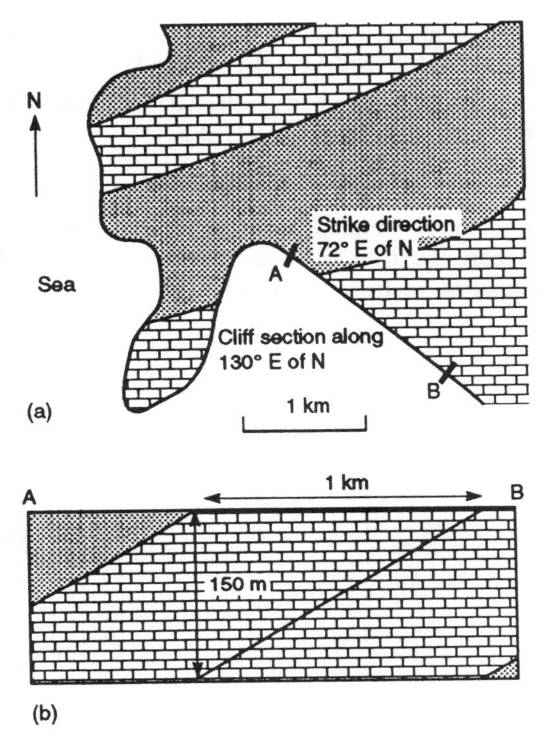

Figure 5.20 (a) Geological map showing, in particular, the position and direction of alignment of the section shown in (b). The map also indicates the direction of strike for the beds outcropping in the section.

FURTHER QUESTIONS

5.17 Evaluate:
(i) $\cos(15°)$, (ii) $\sin(1.2\text{ radians})$, (iii) $\tan^{-1}(0.5)$, (iv) $\cos^2(27°)$, (v) $(\tan(0.5°))^{-1}$.

5.18 Assuming that the earth is a perfect sphere, show that the radius, r, of a circle of latitude is given by

$$r = R\cos(\phi)$$

where R is the earth's radius and ϕ is the latitude.

5.19 An alluvial fan slopes at an angle of 5° to the horizontal and the distance from the fan origin to its base (measured along the fan surface) is 5 km. Calculate the vertical height of the fan origin above its base.

5.20 Look at Figure 5.20 which shows a geological map (Figure 5.20(a)) and a section (Figure 5.20(b)) drawn from a cliff at the location shown on the map. The map indicates that in the area of the section the direction of strike is 72° E of N and the section itself is aligned along 130° E of N. From the information shown on the section, determine the true dip of the beds.

5.21 Remanent magnetism of 10 specimens collected from a Tertiary sill had the following azimuthal directions in degrees E of N:

331, 5, 347, 351, 3, 342, 338, 355, 349, 17.

Assuming that the remanent magnetism strengths are equal (say unity), calculate the vector mean direction for these measurements. Do this either graphically or by calculating vector components.

More about graphs

6

6.1 INTRODUCTION

In one sense geologists frequently have too many data. A field geologist may have a notebook full of dip, strike and location measurements, a geochemist may have analyses of 10 different elements in 100 different rock samples or a geophysicist may have more than a kilometre of computer tape for every kilometre of a 1000 km seismic survey. In all these cases the problem is the same; the scientist involved must somehow make sense of a mass of data far too large to be digested raw. There are three things that can be done about this:

1. Throw away most of the data. Usually this means ignoring all data which do not fit some preconceived notion. This is very definitely not recommended although it is quite frequently done!
2. Perform a statistical analysis. This is the subject of Chapter 7.
3. Plot the data on a graph which will allow the general properties of the data to be visualized. This is the subject of this chapter.

In fact, although I have separated them here, statistics and graphing are subjects which overlap very significantly.

This chapter deals with graphs in which each data item is plotted as a point on a suitable piece of graph paper. The most common graph of this type has already been used extensively, particularly in Chapter 2. This is the simple x–y graph which has two axes, at right angles to each other, representing two different quantities. Figure 6.1 shows such a graph which plots sediment density against depth in a well. Each point represents a specific measurement of depth and density. The remainder of this chapter is about variations upon this simple theme.

6.2 LOG-NORMAL AND LOG-LOG GRAPHS

The use of logarithms, to enable a wide spread of data to be visualized, has already been introduced in Chapter 2. Table 6.1 gives the masses of various

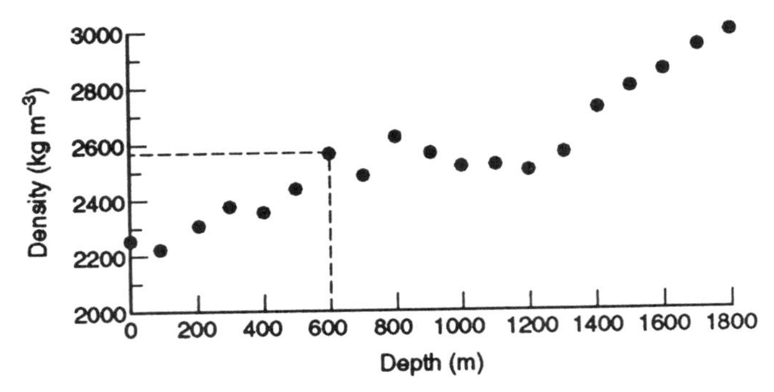

Figure 6.1 Example of a simple x–y plot showing how sediment density varies with depth in a particular well. Each point on this graph represents a specific measurement of density at a particular depth, e.g. at a depth of 600 m the density is 2560 kg m^{-3}.

Table 6.1 Masses and total area of foot soles for various modern and extinct animals

Animal	*Mass (kg)*	*Foot area (m^2)*	*Log(mass)*
Apatosaurus	35 000	1.2	4.54
Tyrannosaurus	7 000	0.6	3.85
Iguanodon	5 000	0.4	3.70
African elephant	4 500	0.6	3.65
Cow	600	0.04	2.78
Human	70	0.035	1.85

Data taken from Alexander, R. (1989) *Dynamics of Dinosaurs and other Extinct Giants*, Columbia University Press, New York.

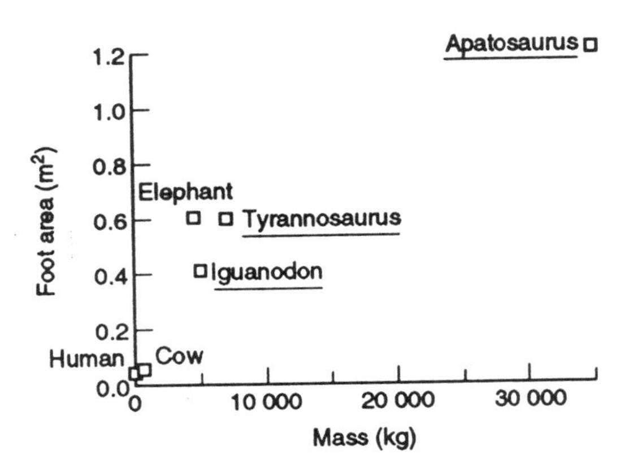

Figure 6.2 Simple x–y plot of the data from Table 6.1. Note that five out of six data points are squeezed into the leftmost $\frac{1}{5}$ of the graph.

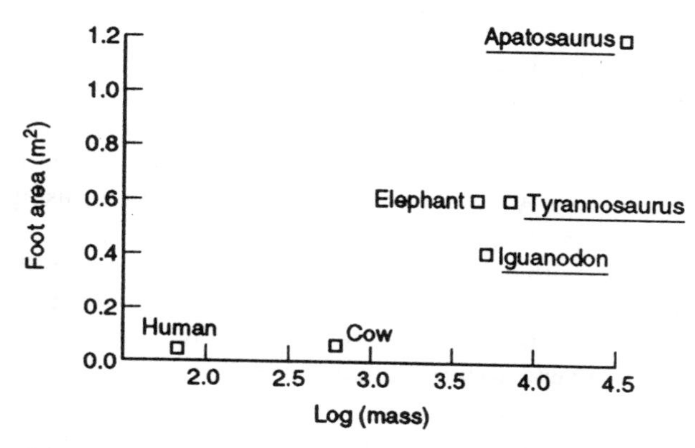

Figure 6.3 Plot of foot area as a function of the logarithm of the mass. This allows the data to be more easily viewed since they are now more evenly spread across the plot.

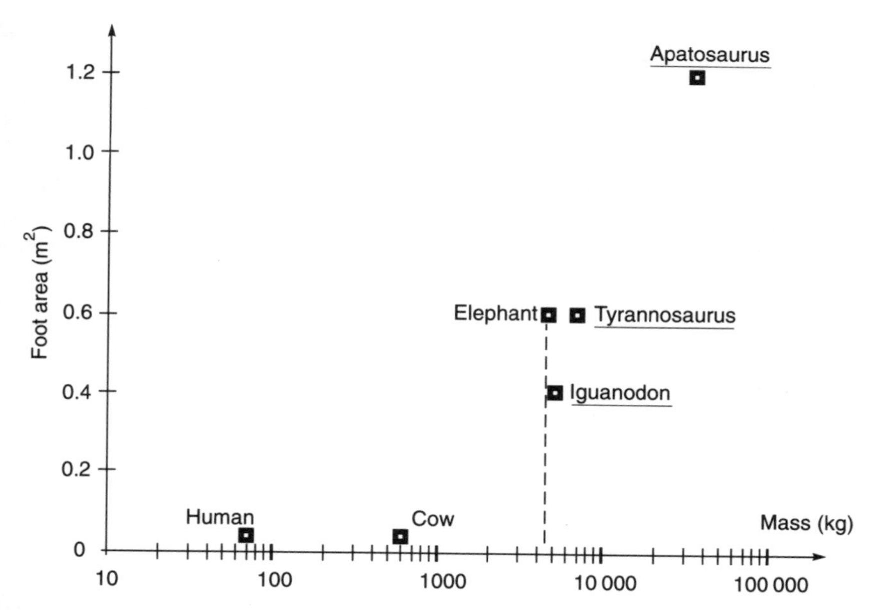

Figure 6.4 Similar to Figure 6.3 except that, instead of plotting the logarithm of the mass, a logarithmically scaled axis has been used. Note that, for example, the distance from 10 to 100 kg is the same as from 100 to 1000 kg and also 1000 to 10 000 kg. The tick marks between 1000 and 10 000 kg are at 1000 kg intervals. The elephant mass is therefore around 4500 kg.

modern and extinct animals together with the total areas of the soles of their feet (this is relevant to whether these animals could walk on soft mud without sinking in and can help to indicate the environment they lived in). These data are plotted on an x–y type graph in Figure 6.2. Note that all the points except one are squeezed into the leftmost fifth of this graph which makes the graph difficult to analyse. A solution to this problem is to plot using the logarithm of the mass instead (Figure 6.3).

The problem with Figure 6.3, however, is that it is now difficult to read off values on the horizontal axis. For example, without looking at Table 6.1, what is the mass of an elephant? You have to read down to the axis (gives 3.65) and then take the inverse logarithm (i.e. mass $= 10^{3.65} = 4467$ kg). This is rather tedious and error prone. An alternative, shown in Figure 6.4, is to use a logarithmically scaled axis. Note that the distance on the horizontal axis between 100 and 1000 kg (a tenfold increase) is the same as the distance between 10 and 100 kg (also a tenfold increase). The result is a graph whose shape is identical to that of Figure 6.3 but from which it is much easier to read the mass of any given animal. Such a graph is known as a **log-normal** plot since one axis (the horizontal one in this case) is scaled logarithmically whilst the other has a normal scale. The mass of the elephant can now be read off the axis as around 4500 kg. The oddly arranged tick marks on this axis will be explained presently.

Question 6.1 Table 6.2 gives the average frequency (number per year) of earthquakes of various magnitudes over the period 1918–1945. Using these data:

(i) Plot the frequency as a function of magnitude on normal graph paper. (Frequency should be on the vertical axis.)
(ii) Plot the frequency as a function of magnitude on log-normal graph paper.

Table 6.2 Average earthquake frequency, as a function of magnitude, between 1918 and 1945

Magnitude	*Number per year*
8	1
7	18
6	108
5	800
4	6 200
3	49 000
2	300 000

Data from Gutenberg, B. and Richter, C. F. (1954) *Seismicity of the Earth and Associated Phenomena*, Princeton University Press, Princeton.

(iii) Plot log(frequency) against magnitude on normal graph paper. Estimate the constant b in the equation $\text{Log}\, N = k - bM$ where N is frequency, M is magnitude and k is a constant.

It would have been more difficult, although not impossible, to estimate b from the second graph you plotted. Thus, if the main objective is to display the data more clearly, use log-normal graph paper but, if the objective is to estimate a parameter such as b, take logarithms first and plot on normal graph paper (cf. section 3.5).

In Table 6.1, the foot areas are also spread over a rather large range. It might be useful, therefore, to take logarithms of the areas or to use logarithmically scaled axes in both directions (Figure 6.5). Figure 6.5 is an example of a **log-log** plot. Note that Figure 6.5 includes a logarithmically scaled grid. A grid like this is usually drawn on logarithmic graph paper and enables intermediate values to be read more easily. The way in which these are used is indicated on Figure 6.5 where it can be seen that the *Iguanodon* has a mass of 5000 kg and a total foot area of 0.4 m^2. Note that the size interval between each line is not fixed. Between 1000 and 10 000 kg the interval is 1000 kg (as shown by the *Iguanodon* example) but between 10 000 and 100 000 kg the interval becomes 10 000 kg. Similarly, on the vertical axis, the lines are 0.1 m^2 apart between 0.1 and 1 m^2 whilst they are 1.0 m^2 apart between 1 and 10 m^2.

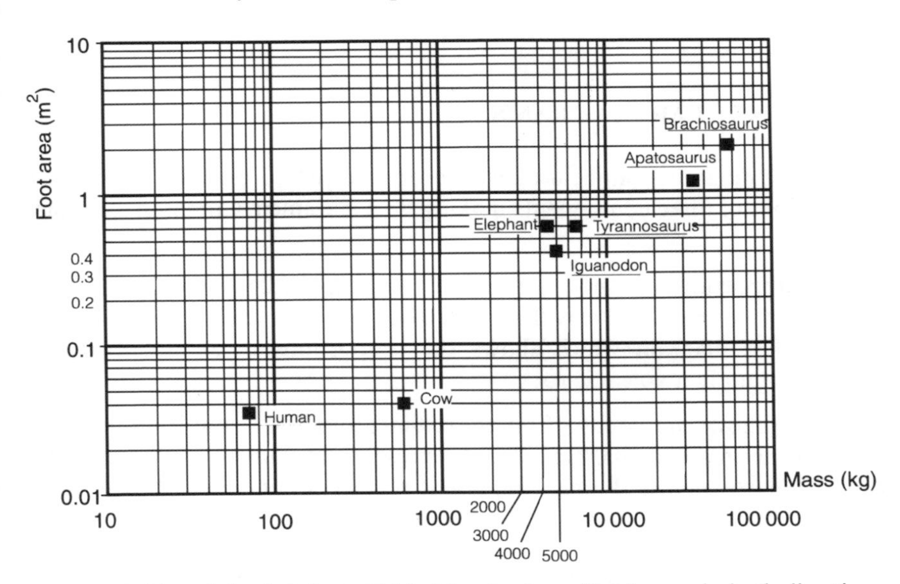

Figure 6.5 Plot of the data from Table 6.1 using logarithmic axes in both directions. Note the inclusion of a logarithmically scaled grid. Using this, the *Iguanodon* can be seen to have a mass of 5000 kg and a foot area of 0.4 m^2.

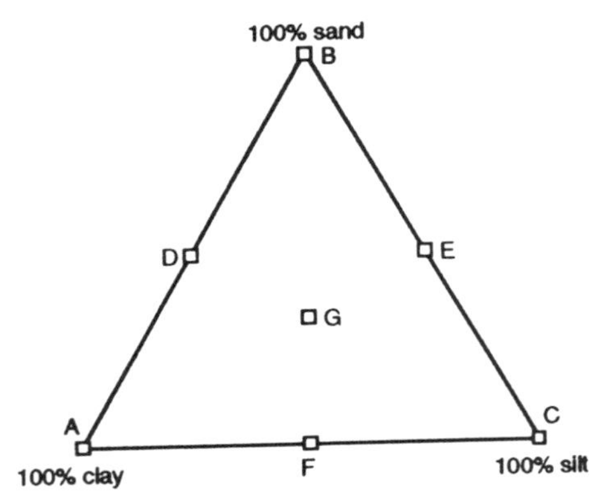

Figure 6.6 A triangular diagram illustrating the proportions of clay, sand and silt in different sedimentary rock samples. The corner points, A, B and C represent samples consisting of pure clay, sand or silt respectively. Points D, E and F are 50:50 mixtures. Point G corresponds to a sample containing equal amounts of clay, sand and silt.

Question 6.2 Using Figure 6.5, what is the mass and total foot area of *Brachiosaurus*?

Incidentally, in this example using a logarithmic scale for the vertical axis has only made a marginal improvement. However, in other cases it will make a much more useful alteration in the distribution of the data across the graph.

6.3 TRIANGULAR DIAGRAMS

Triangular diagrams can be used whenever you wish to visualize the relative proportions of three components making up a specimen. Common examples are:

1. The proportions of sand (particles between 2 and 0.063 mm diameter), silt (0.063–0.004 mm) and clay (less than 0.004 mm) in a sedimentary rock.
2. An **AFM diagram** which shows the proportions of alkalis, iron and magnesium in a volcanic rock.

Figure 6.6 shows simple cases from the sedimentological example. Point A sits in the corner marked '100% clay' and represents a sediment containing only clay. Similarly, points B and C represent rocks containing exclusively sand and silt respectively. Point D lies half-way along a line joining 100% clay to 100% sand. It represents a sediment half of which is clay and half of which is sand. Similarly point E is a 50:50 sand–silt mixture whilst point F is a 50:50 clay–silt mixture. Finally, point G is in the centre of the triangle, equidistant from all three edges, and represents a sediment which consists of a $\frac{1}{3}$ clay, $\frac{1}{3}$ sand and $\frac{1}{3}$ silt mixture.

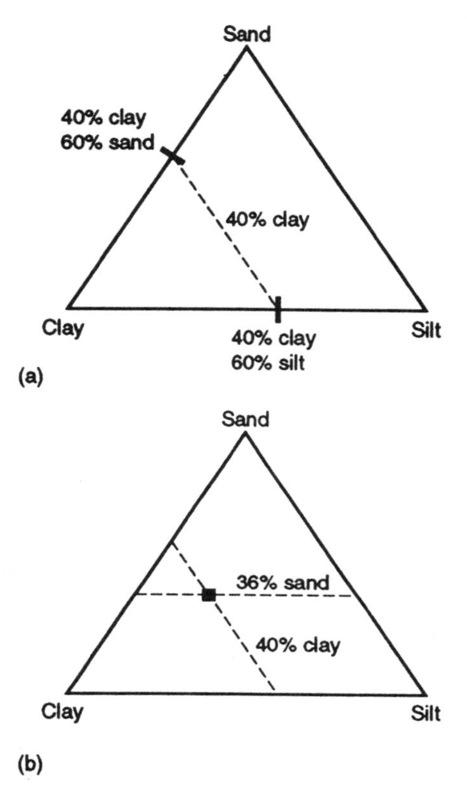

Figure 6.7 Plotting a point which is 40% clay, 24% silt and 36% sand. (a) The dashed line represents all points which are 40% clay and is constructed by joining the point which is 40% clay–60% sand to the point which is 40% clay–60% silt. (b) The location for the point is found by constructing a similar line for all points which are 36% sand and finding where this intersects the 40% clay line.

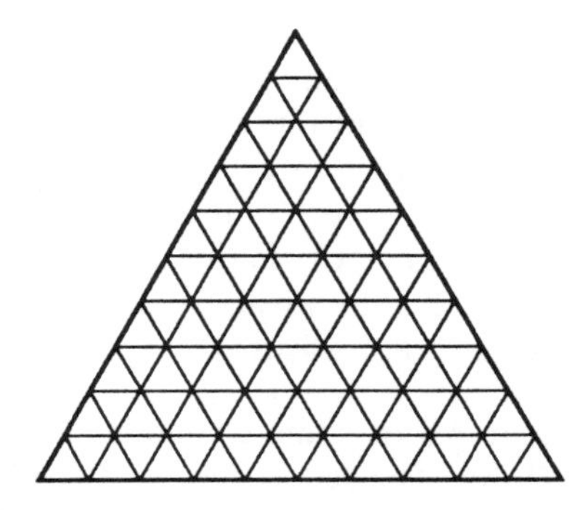

Figure 6.8 An example of graph paper used for plotting triangular diagrams. In this example, lines have been drawn at 10% intervals although such graph paper will normally be marked at 1% intervals.

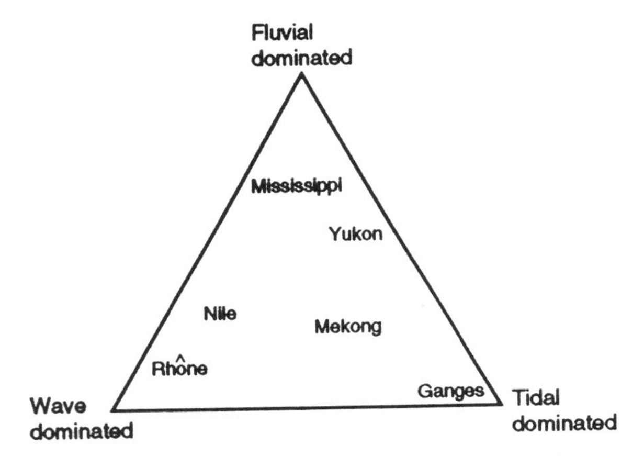

Figure 6.9 Use of a triangular diagram for classification of delta types. Deltas are rarely completely dominated by wave, tidal or fluvial processes but are controlled by a combination of these.

It is also, I think, fairly obvious where to plot a point corresponding to, say, 40% clay and 60% sand and which therefore contains no silt. This point will be on the line joining 100% sand to 100% clay and will be 40% of the distance along from sand to clay (or, equivalently, 60% of the distance along from clay to sand), i.e. slightly closer to sand than to clay.

Question 6.3 Plot the 40% clay, 60% sand point on to Figure 6.6.

These examples are relatively straightforward and it is quite easy to see where on the triangular plot each of these points should go. What about a sediment containing 36% sand, 24% silt and 40% clay? Figure 6.7 illustrates how this is done. In Figure 6.7(a) a line has been drawn which connects the 40% clay, 60% sand point to a point representing 40% clay, 60% silt. All points along this line contain 40% clay but have differing amounts of sand and silt making up the remaining 60%. Similarly, Figure 6.7(b) shows a line representing all points which have 36% sand. The point where the 40% clay line intersects the 36% sand line is, of course, a point representing a sediment with 36% sand and 40% clay and which must, therefore, be 24% silt.

To assist in accurate plotting of such points, a triangular net similar to that shown in Figure 6.8 is used. For clarity in this illustration, the lines are drawn at 10% intervals although these lines will usually be plotted at 1% intervals on most sheets of triangular graph paper.

Question 6.4 Use Figure 6.8 or some triangular graph paper to plot an AFM diagram as follows. The left corner of the plot represents 100% ($Na_2O + K_2O$). The right-hand corner represents 100% MgO. The top corner represents 100% ($FeO + Fe_2O_3$). Mark these points on your graph and then plot the following data which are taken from a set of related volcanic rocks:

(a) 10% ($Na_2O + K_2O$), 45% MgO, 45% ($FeO + Fe_2O_3$).
(b) 10% ($Na_2O + K_2O$), 35% MgO, 55% ($FeO + Fe_2O_3$).
(c) 10% ($Na_2O + K_2O$), 25% MgO, 65% ($FeO + Fe_2O_3$).
(d) 12% ($Na_2O + K_2O$), 20% MgO, 68% ($FeO + Fe_2O_3$).
(e) 15% ($Na_2O + K_2O$), 15% MgO, 70% ($FeO + Fe_2O_3$).
(f) 18% ($Na_2O + K_2O$), 12% MgO, 70% ($FeO + Fe_2O_3$).
(g) 23% ($Na_2O + K_2O$), 12% MgO, 65% ($FeO + Fe_2O_3$).

A graph such as this can furnish significant information about the evolution of a volcanic rock series. However, the way in which this is done, as well as the details of how to obtain the numbers to plot, is beyond the scope of this book.

Before leaving the subject of triangular diagrams, it is worth mentioning that they can be used for classification of geological features if there are three clear **end members** to such a classification scheme. For example, deltas are commonly classified as being dominated by fluvial, wave or tidal processes. However, real deltas are influenced to some extent by all three types of process and will not be accurately represented by a simple threefold classification scheme. The solution is to use a triangular diagram to represent all possible deltas (Figure 6.9). Real deltas will then fall at some point within the diagram which represents the proportions of fluvial, wave and tidal effects governing the geometry.

6.4 POLAR GRAPHS

Some types of data are naturally cyclic. For example, the data in Table 6.3 give the strength of the non-dipole portion of the earth's magnetic field at

Table 6.3 Non-dipole magnetic strength (in micro tesla) at various locations on the earth's equator

Longitude (degrees E)	*Non-dipole strength (μT)*
0	17.5
30	13
60	6
90	9.5
120	9.5
150	7
180	4.5
210	3
240	3
270	2
300	5
330	14

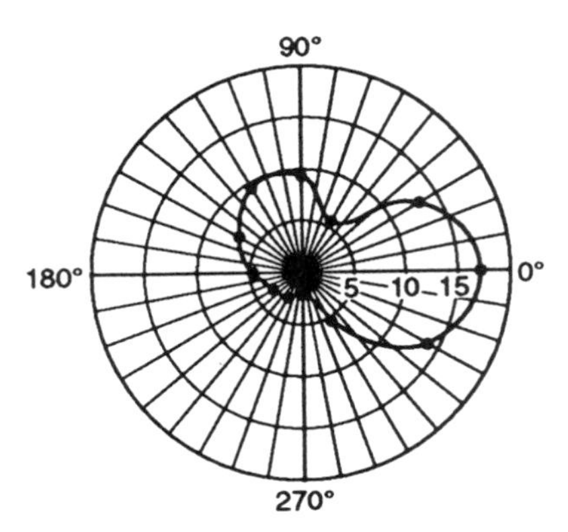

Figure 6.10 The non-dipole field data from Table 6.3 plotted in polar form to emphasize its cyclic nature.

various locations around the equator (the non-dipole field is that part of the magnetic field which cannot be explained as due to a simple bar magnet). Now, the data at a longitude of 330° are only 30° from the data at 0° but, on an $x-y$ plot, they would appear at the opposite end of the graph. Using a polar plot avoids this problem (Figure 6.10). In the polar plot, the longitude is plotted around the circumference of a circle whilst the field strength is given by the distance from the plot centre (i.e. the stronger the field the further the point is from the centre).

6.5 EQUAL INTERVAL, EQUAL ANGLE AND EQUAL AREA PROJECTIONS OF A SPHERE

This section deals with the problem of plotting data measured on the surface of a sphere on to a flat sheet of paper. This problem occurs, for example, in map making when it is necessary to represent a large portion of the earth's surface by a map in an atlas. This cannot be done without distortion, and there are therefore a large number of different ways of doing this each of which has advantages and disadvantages. This section will deal with three very similar methods which are widely used in structural geology, crystallography, earthquake seismology and many other branches of earch science. These **projections** are useful whenever information about directions in three dimensions is plotted and they enable many, otherwise complex, manipulations to be carried out relatively simply. There are subtle differences between the methods used in different branches of geology but there is a core of ideas and methods which is common throughout. In this section I introduce some of these ideas but, it must be emphasized, application in particular fields has much more extensive uses than those described here. This section is very much

Table 6.4 Apparent dip measurements from a single bed at eight separate locations. The azimuth is the direction in which each of the dip measurements was taken

Location	*Apparent dip (°)*	*Azimuth (°E of N)*
1	44	11
2	12	305
3	31	79
4	42	2
5	21	318
6	34	337
7	7	112
8	39	352

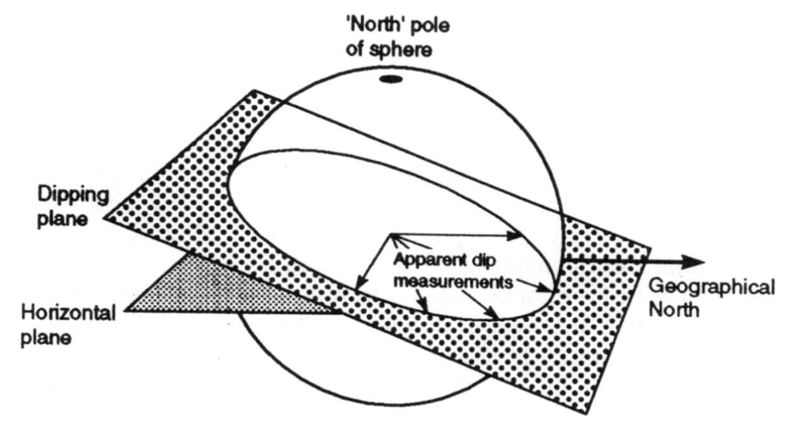

Figure 6.11 Spherical projection of apparent dip measurements. These will lie on a great circle if the bed is planar.

a starting point for the more detailed discussions you will meet in specific geological sub-disciplines.

Consider a bed which outcrops at various locations and whose apparent dip has been measured along a different orientation at each of the different locations. Table 6.4 lists such a series of apparent dip measurements together with the directions in which these dips were measured. Is the bed a simple planar dipping one or is the bed folded in some way? If the bed is planar, what is the true dip and dip direction?

Figure 6.11 illustrates the starting point for resolving these issues. This diagram assumes that the bed is indeed a simple dipping planar bed represented by the dipping plane in this figure. The arrows drawn on the plane represent measurements of apparent dip of this surface taken in various orientations. A sphere is drawn with its centre on the plane. It is useful to think of this sphere as having a 'north' pole at its top and a 'south' pole at the bottom but be careful not to confuse this meaning of north with the direction of geographical north. The intersection of the plane and sphere is a **great circle**. A great circle is any circle on the surface of a sphere whose centre lies at the centre of the sphere. Thus, the equator and lines of longitude on the earth's surface are great circles but lines of latitude are not. The individual apparent dip measurements start at the centre of the sphere and intersect the sphere at points on the great circle. If you think of the sphere as having lines of latitude and longitude in a similar fashion to the earth, each point plots at a 'southerly' latitude equal to the apparent dip and at a longitude equal to the azimuth. Thus, if we plot the projection of our dip measurements on to a sphere and if the resulting points lie along a great circle, the measured bed is a simple dipping plane. On the other hand, if this **spherical projection** of the dip measurements is not a great circle, the bed is not planar. Note that, because dips are always measured below the horizontal, only half of all possible orientations are represented. Thus, the spherical projection of the dip data should actually define a semicircle (i.e. the lower half of the great circle).

The problem now is that plotting and performing measurements on a sphere is not very convenient. A solution would be to plot the data using a projection which represents the surface of the sphere on a flat sheet of graph paper. A method for doing this, bearing in mind that we only have to deal with the lower hemisphere, is to plot on to polar graph paper. To do this, the azimuth is plotted around the circumference of the plot and distance, r, from the centre of the plot is used to represent distance from the south pole of the sphere. There are, however, many ways of doing this. The simplest is just to let this distance be proportional to the angular distance from the pole. Thus, if a point on the plot represents a dip of ϕ, the corresponding distance, r, on the plot is given by

$$r = R(90 - \phi)/90 \tag{6.1}$$

where R is the plot radius. In other words, dip increases linearly from zero at the plot circumference to 90° at the plot centre. For example, a dip of 45° would plot at a distance of

$$\begin{aligned} r &= R(90 - 45)/90 \\ &= R/2 \end{aligned}$$

i.e. half-way between the plot edge and the plot centre. Similarly, a dip of 0° would be at the plot edge (i.e. $r = R$) and a dip of 90° would be at the graph centre (i.e. $r = 0$).

Question 6.5 Using equation (6.1), calculate how far apart two points with the same azimuth but dips of 10^0 and 20° are. Assume the graph radius, R, is 10 cm. Repeat this calculation for dips of 70° and 80°.

Using this method for plotting the measurements in Table 6.4 results in Figure 6.12 and is known as an **equal interval projection**. The data certainly

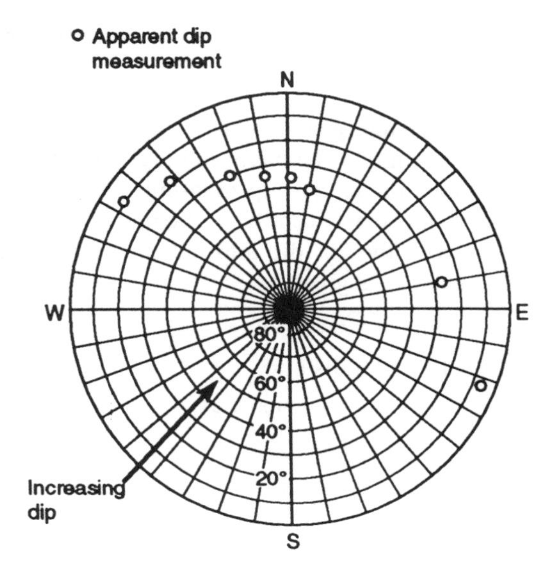

Figure 6.12 Equal interval polar plot of the dip azimuth data in Table 6.4.

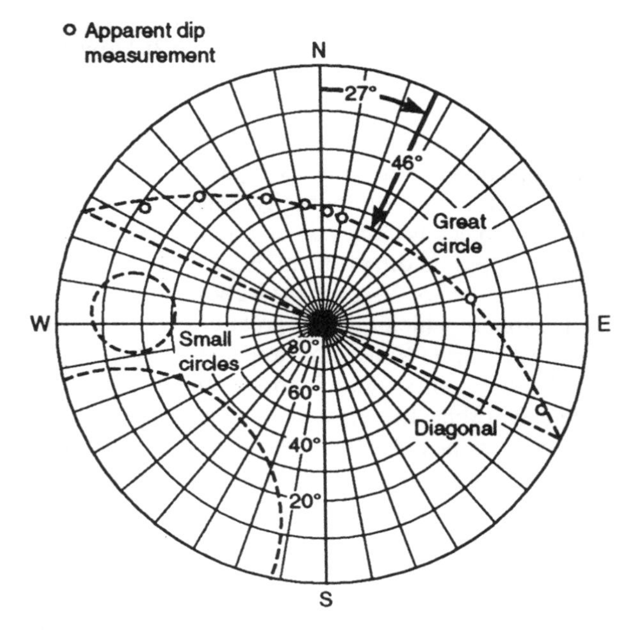

Figure 6.13 Stereographic (equal angle) projection of apparent dip data. A great circle is drawn through these points and the position where this circle has maximum dip (point A) corresponds to a dip of 46° in a direction 27° E of N. Two examples of small circles are also shown.

look as if they might lie along a semicircle but we have the problem that we do not really know whether this is half of a great circle or just some other fairly smooth curve.

Figure 6.13 shows a **stereographic** or **equal angle projection** of the apparent dip data. This is very similar to the equal interval plot (Figure 6.12) except that the distance between the circles representing dip is not constant. Note, for example, that the distance between the 0° dip and 10° dip circles is greater than the distance between the 70° dip and 80° dip circles. In this case, the distance from the plot centre is given by

$$r = R\tan[(90 - \phi)/2]. \qquad (6.2)$$

Thus, a dip of 45° would plot a distance

$$\begin{aligned} r &= R\tan[(90-45)/2] \\ &= R\tan(22.5) \\ &= 0.414R \end{aligned}$$

which is significantly closer to the centre than half-way out (i.e. closer to the centre than in the equal interval plot). Applying equation (6.2) to the cases of zero dip and 90° dip gives a result of $r = R$ and $r = 0$ respectively (the same as for the equal interval plot).

Question 6.6 Repeat question 6.5 using equation (6.2). How do these results compare to the equal interval case?

The equal angle projection has two important properties. Firstly, angles measured on the projection are the same as angles on the surface of the sphere. This is particularly useful in crystallography. Secondly, circles drawn on the surface of the sphere project as circles on an equal angle plot and this will be useful for solving our apparent dip problem. Figure 6.13 shows three examples of projections of circles. Two of these are projections of **small circles** (i.e. circles on the surface of a sphere which are not great circles). A great circle is also plotted and is an arc of a circle whose start and end locations define a diagonal to the plot since the two points where it crosses the 0° dip line on the sphere must be opposite each other. In particular, the great circle plotted has been chosen to pass as close as possible through the dip/azimuth data.

This great circle clearly passes quite well through the data points so we can say that the apparent dip measurements are indeed taken from a simple planar dipping bed. The true dip and dip direction can also now be found. A dip measurement made in any direction other than the true direction of dip must be smaller than the true dip. Hence, the true dip corresponds to the maximum dip crossed by the great circle. This occurs at the point marked A and the true dip and dip direction can then be read off as 46° in a direction 27° E of N.

The equal angle projection has the disadvantage that it distorts areas. A figure near the circumference of the plot will actually project as an area four times larger than an identical figure at the plot centre. Figure 6.14 shows an

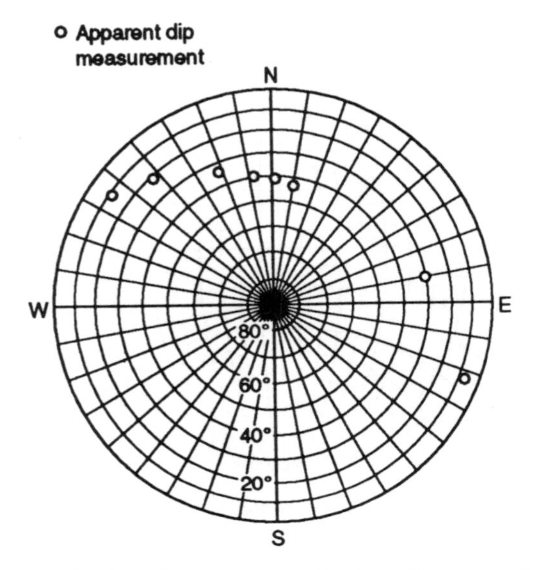

Figure 6.14 Equal area projection of the apparent dip data. Note that the concentric circles now get closer together towards the edge of the plot.

equal area projection of the apparent dip data. In this case the concentric circles get closer together towards the edge of the plot and r is given by

$$r = \sqrt{(2)}R\sin[(90 - \phi)/2]. \tag{6.3}$$

This time, a dip of 45° will plot at a distance

$$\begin{aligned} r &= \sqrt{(2)}R\sin[(90 - 45)/2] \\ &= \sqrt{(2)}R\sin(22.5) \\ &= 0.541R \end{aligned}$$

which is significantly further from the centre than half-way out (i.e. further out than in the equal interval plot). Applying equation (6.3) to the cases of zero dip and 90° dip gives a result of $r = R$ and $r = 0$ respectively (the same as for the equal interval plot).

Question 6.7 Repeat question 6.5 using equation (6.3). This time compare your results to both the equal interval and equal angle plots

The equal area plot has the property that equal areas on the surface of the sphere project as equal areas on the plot. The disadvantage is that circles and angles are now distorted. The equal area projection is frequently used in structural problems rather than the equal angle projection since the density of plotted points is often important and this is distorted by the equal angle projection.

To make it easier to find great circles (and indeed small circles) on these projections, a slightly different type of display from the polar plots is normally

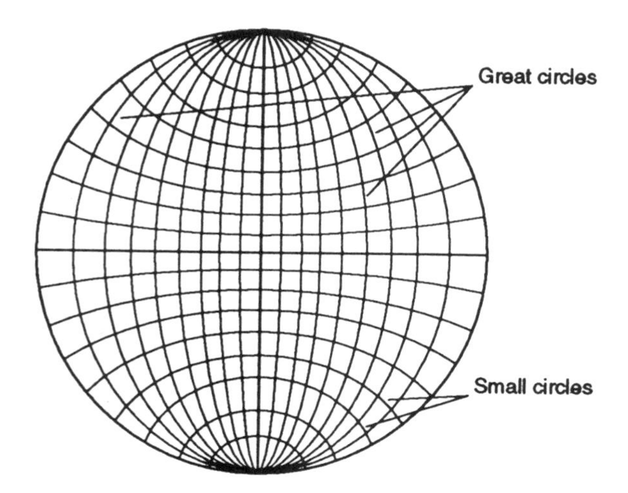

Figure 6.15 Wulff stereographic net which has great circles and small circles already plotted on it.

used. These are called **equatorial nets** and are drawn with great circles and small circles already plotted upon them. Figure 6.15 shows the equatorial net for the equal angle case. This is known as a **Wulff net**. The equivalent plots for the equal interval and equal area projections are called **Kavraiskii nets** and **Schmidt nets** respectively and have a similar appearance to the Wulff net. In Figure 6.15, the great circles and small circles cut the vertical and horizontal axes at 10° intervals and cut the circumference of the net at 10° azimuth intervals (these lines will usually be drawn at even finer intervals on these nets).

These nets are extremely useful but plotting a point on an equatorial net is slightly more involved than on the polar type plots. Plotting the position of data points is normally achieved using an equatorial net mounted on a board with a drawing pin through the centre of the net and into a piece of tracing paper placed over the net. This arrangement allows the tracing paper, upon which the data are plotted, to be rotated above the net. Figures 6.16(a)–(c) show how to plot one of the data points from Table 6.4 (location 3, dip 31°, azimuth 79°) as follows. Figure 6.16(a) shows a Wulff net with north marked on to the tracing paper at the top of the net. In Figure 6.16(b) the tracing paper is rotated anticlockwise by 79° and the point is plotted 31° down from the top of the net. Rotating the tracing paper so that north is again at the top results in the point being correctly positioned over the net (Figure 6.16(c)).

Repeating this procedure for all of the data points from Table 6.4 results in Figure 6.16(d). This figure can then be rotated until the points lie along a great circle (Figure 6.16(e)). The distance of this great circle from the plot edge gives the maximum dip value (46°). Finally, rotating the paper again so that north is uppermost results in Figure 6.16(f) in which the data and the great circle appear in their correct positions and it can be seen that the dip direction

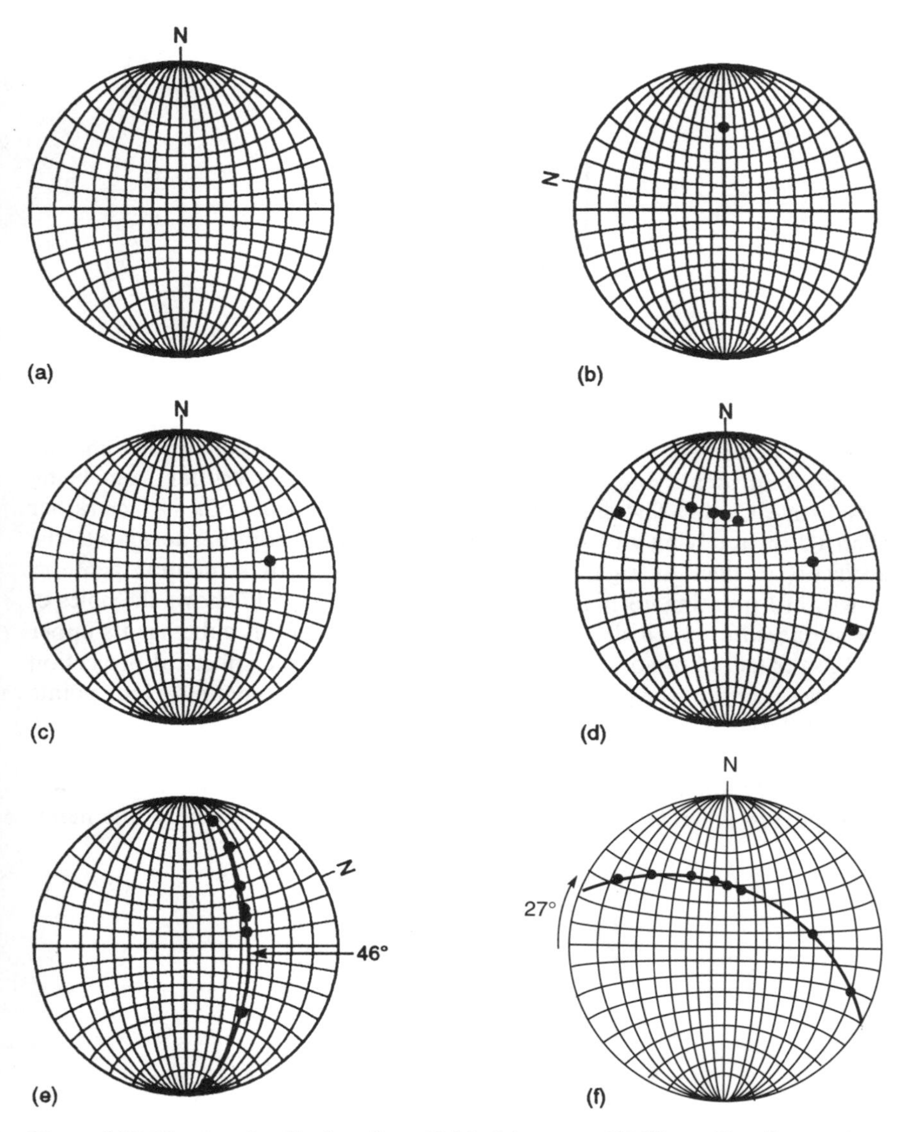

Figure 6.16 Plotting the dip data from Table 6.4 on to a Wulff net. Details are given in the text.

is 27° E of N. Exactly the same set of manipulations can be performed with either the Kavraiskii or the Schmidt nets and, indeed, for many purposes these nets are interchangeable.

The above is only one example of the many problems which these projections can solve. Other examples are: for determining the axes of folded structures; determining the earlier orientations of structures which have been

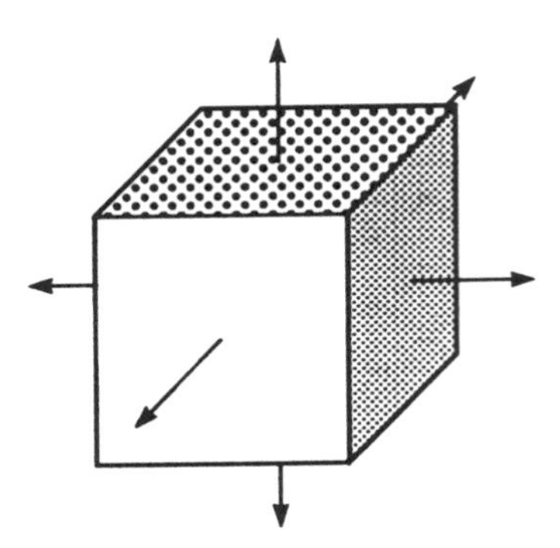

Figure 6.17 The poles of the six faces of a cube.

multiply deformed; determining detailed earthquake mechanisms; characterizing and identifying crystal structures. This list is very far from being exhaustive. Quite a few of these involve plotting the **poles** of a surface rather than a direction lying on a surface. Poles are outward-pointing **normals** to the surface (i.e. lines at right angles to the surface). In structural geology, downward-pointing normals are used instead. Figure 6.17 shows the poles of the surfaces forming a simple cube. As you can see, if this cube has its upper and lower faces horizontal, these poles would plot in a spherical projection with two points at the poles of the sphere and with the remaining four points around the equator.

A small complication with transferring these data to a stereographic (or other) projection is that we have points plotted in both the upper and lower hemispheres, whereas a stereographic projection represents only one hemi-

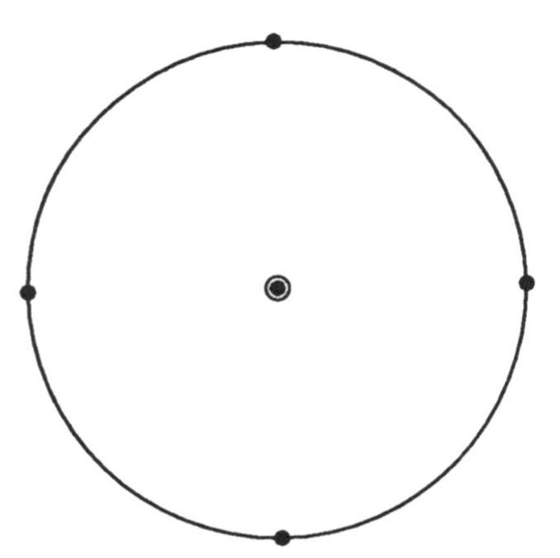

Figure 6.18 Stereographic projection of cube poles. The top and bottom faces plot as the dot and circle respectively in the centre. The other four poles plot as the dots around the circumference.

sphere. The solution is to use two projections, one for the upper hemisphere and one for the lower. In practice, both sets of points are plotted on one graph and the difference between them is indicated by using dots for the upper hemisphere points and circles for the lower hemisphere points. Thus, the stereographic projection of the cube poles produces Figure 6.18.

FURTHER QUESTIONS

6.8 Sedimentary beds, when folded, can have three types of geometry:

(i) planar (i.e. not folded at all);
(ii) cylindrical (i.e. folded around one axis, think of a towel hanging over a towel rail);
(iii) isoclinal (i.e. dome shaped).

In practice, these are the extreme types of fold and real beds are deformed using a combination of these. Thus, for example, a bed might have a very gentle cylindrical fold which can be thought of as a combination of the planar and cylindrical end members. Another bed might be tightly folded around one axis and be gently folded about another axis at right angles (imagine the towel on the towel rail again but this time the rail itself is bent up in the middle). This would be a combination of cylindrical and isoclinal folding.

What type of graph would be suitable for illustrating the above concepts?

6.9 Table 6.5 shows the total area of continental crust which is older than a given age. Plot these data in a variety of ways and decide which, you believe, shows the data best.

Table 6.5 Area of continental basement which is older than a given age

Age (My)	*Area (10^6 km^2)*
>450	91.1
>900	50.0
>1350	35.4
>1800	26.7
>2250	7.3
>2700	1.1

Source: Hurley, P. and Rand, J. (1969). Pre-drift continental nuclei. *Science*, **164**, 1229–42.

6.10 Cross-sections through the earth taken parallel to the equator are, to a good approximation, circular, However, there are small variations due to mountain ranges, ocean basins, etc. Imagine such a section taken at a latitude of, say, 30° N. The difference between a circle and the actual section could be tabulated as a function of longitude. What form of plot would be best for displaying such data?

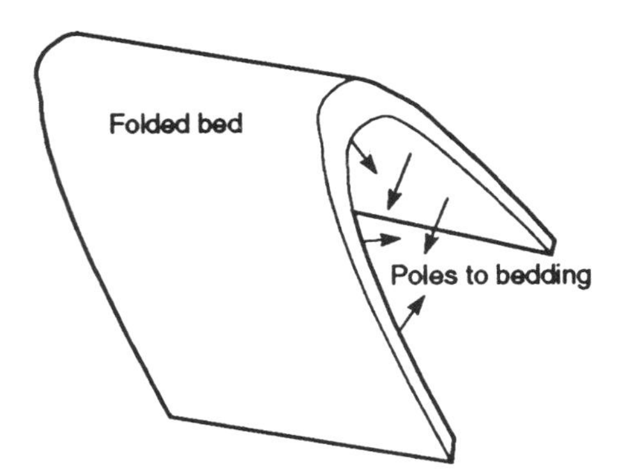

Figure 6.19 A sedimentary bed has been folded around a single axis as shown. The poles to this bed are also shown. Note that the poles all lie in the same plane but point in different directions depending upon which part of the bed they sit on.

6.11 Consider Figure 6.19 which shows a planar bed which has been folded about a single axis. Some of the poles to the bedding are also shown and these point in various directions because of the fold.

(i) What form would you expect a spherical projection of the poles to bedding to take?

(ii) How does the direction and dip of a pole relate to the direction and amount of bed dip at any particular point? For example, if a bed dips towards 30° E of N what direction does the pole point? If the same bed has a dip of 20° what is the dip of the pole?

(iii) Determine the poles resulting from the following bed dip data and plot them on to an equatorial net:

Dip (°)	*Dip direction (°E of N)*
70	182
65	170
25	131
40	70
36	90
70	35
40	40
37	73
37	146

(Data from McClay, K. (1987) *The Mapping of Geological Structures*, John Wiley, Chichester.)

(iv) Is this information consistent with being taken from a bed folded around one axis?

Statistics

7

7.1 INTRODUCTION

This chapter is a very brief introduction to the subject of geological statistics. Statistics is probably the most intensively used branch of mathematics in the earth sciences. For this reason, even an introduction to the subject fills an entire book and there is a large number of such texts. I do not intend, therefore, to cover this topic in the depth it deserves but to give an introduction which I hope will help ease you into the subject and allow you to go on to other texts with some idea of what to expect.

A major problem with statistics is that it is very easy to mislead. A good example comes from the statisticians' favourite subject, coin-tossing. If a coin is tossed six times it is quite likely that there will be three heads. It is very unlikely that six heads will occur. If I then went on to state that it is more probable that the result will be HTHTHT than HHHHHH (where H represents heads and T tails) I would be seriously misleading you. Both of these events are equally unlikely! The reason that the most likely result is three heads and three tails is that there is a large number of ways of doing this (e.g. HHTHTT, HTHTTH and HHHTTT) whereas there is only one way to get six heads (i.e. HHHHHH). Any particular combination of heads and tails is as likely as any other. Other ways in which statistics can mislead are much more subtle and even experts can, and do, make very serious errors. However, do not let me put you off statistics! If you work carefully and thoughtfully statistics can produce results that could not be obtained in any other way.

Question 7.1 Write down all possible results of tossing a coin three times. Tabulate the results in terms of the number of different ways of obtaining 0 heads, 1 head, 2 heads or 3 heads

7.2 WHAT IS A STATISTIC?

I have been writing up to now as if everybody knows exactly what a statistic is. However, even this term is popularly misused. A statement such as 'in 1970 the oil refining capacity of Belgium was 32.6 million tonnes per year' is a fact, not a statistic. So what is a statistic? Let me start with an example of a situation in which statistics might be useful.

Consider a pebbly beach. How would you go about determining the typical composition, mass and length of the pebbles on this particular beach? If I were to pick up one pebble from this beach I would have a **specimen** from the beach. This would probably tell me the composition of some of the pebbles. However, this specimen might be very untypical. A better way to get information would be to pick up 100 or 1000 pebbles from random locations on the beach. I would then have a **sample** from the beach. This would give me a much better idea of the most common rocks the pebbles were produced from and their typical masses and lengths. Finally, I could examine (in principle) all the pebbles from the beach. This is the **population** of all pebbles from this beach and I could then make definitive statements about the composition of the beach. To recap, a specimen is one object, a sample is a number of objects and a population is all the relevant objects. Note that the world 'sample' is frequently used in geology to denote a specimen (e.g. 'a sample of sandstone' meaning a single piece of sandstone). This is confusing and I recommend that you use the world 'specimen' whenever possible.

Question 7.2 If I have 6 books from a library containing 10 000 books, do these 6 form a specimen from the library, a sample from the library or the library population?

Now, we can return to the idea of a statistic. Is the average mass of a pebble a statistic? This depends on whether this average is determined from a sample of pebbles or from the total population of pebbles. The average of the population is a **parameter** of the beach and is a simple fact (just like the Belgian oil refining capacity). The average of a sample, on the other hand, is a **statistic**, it is an attempt to estimate the average mass of all the pebbles by calculating the average mass of some of them. In other words, a statistic is an estimate of a parameter based upon a sample of the population. As another example, consider voting patterns in an election. The estimates of voting intentions obtained by polling organizations before the election itself are statistics (they are based on questioning a small minority of voters) whereas the final official result is a parameter of the election.

Returning to the beach example, the way in which the masses vary from pebble to pebble is described by many parameters in addition to the average mass. For example, the pebble masses may all be very close to one another or they may be widely different. One parameter which quantifies this is called the **standard deviation**. This will be defined more precisely in the next section. Another parameter is called the **skew** of the population and this tells us whether there are more pebbles which are heavier than the average or more pebbles which are lighter. All of these parameters would normally be estimated from a sample of the population. Each of the resulting estimates of a parameter is a statistic.

Whether or not these statistics are good estimates depends on how well the sampling was performed and also on the size of the sample. Two pebbles picked up from one place on the beach are unlikely to yield a good estimate of the average mass. One hundred pebbles picked up at random from all over the beach will give a much better estimate. Designing an experiment or fieldwork so that the information collected represents a good sample is an important part of a scientific approach to a problem.

7.3 COMMONLY ENCOUNTERED PARAMETERS AND STATISTICS

Table 7.1 shows the masses of 100 pebbles from a beach. From this sample we might wish to get an idea of the typical mass and also the spread of masses. Each of these can be quantified in several different ways.

Table 7.1 The weights of 100 pebbles sampled from a beach

Pebble No.	*Mass (g)*	*Pebble No.*	*Mass (g)*	*Pebble No.*	*Mass (g)*	*Pebble No.*	*Mass (g)*
1	822	26	375	51	483	76	666
2	355	27	134	52	369	77	871
3	909	28	204	53	496	78	714
4	632	29	161	54	414	79	783
5	706	30	160	55	1480	80	498
6	359	31	419	56	1115	81	539
7	881	32	147	57	618	82	597
8	284	33	68	58	1227	83	1834
9	607	34	91	59	751	84	915
10	1263	35	167	60	1349	85	418
11	290	36	459	61	658	86	191
12	795	37	151	62	360	87	898
13	1120	38	135	63	454	88	224
14	1154	39	80	64	325	89	1473
15	439	40	197	65	1645	90	994
16	229	41	233	66	1243	91	397
17	182	42	75	67	393	92	1158
18	383	43	115	68	648	93	294
19	719	44	314	69	1090	94	1752
20	509	45	414	70	476	95	625
21	322	46	83	71	1017	96	685
22	578	47	239	72	1814	97	1251
23	1336	48	146	73	508	98	499
24	686	49	145	74	636	99	479
25	488	50	62	75	541	100	1331

The typical mass can be described by the **mean**, $\bar{w}$,

$$\bar{w} = (\text{Total mass of the sample})/(\text{Number of pebbles}) \tag{7.1}$$
$$= 61\,018/100 \approx 610\,\text{g}$$

for the pebbles in Table 7.1. Frequently, this computation will be described using the following notation:

$$\bar{w} = \frac{1}{N}\left(\sum_{i=1}^{N} w_i\right) \tag{7.2}$$

where w_i is the mass of the ith pebble and N is the number of pebbles in the sample (i.e. w_1 is the first mass in Table 7.1 (822 g), w_5 is the fifth mass (706 g) and so on). The symbol $\sum$ (sigma) is an instruction to add together the w_i (i.e. add the masses of the pebbles together). The $i = 1$ below the $\sum$ and the N above indicate that all items numbered between 1 and N have to be added (i.e. the mass of 100 pebbles in our example). Finally, the result of this addition is divided by N (i.e. 100 in our case). Equation (7.2) may be rewritten in words as 'the average mass may be found by summing the masses of N pebbles and

Table 7.2 Pebbles ranked according to decreasing weight

Rank	*Mass (g)*	*Rank*	*Mass (g)*	*Rank*	*Mass (g)*	*Rank*	*Mass (g)*
1	1834	26	822	51	496	76	284
2	1814	27	795	52	488	77	239
3	1752	28	783	53	483	78	233
4	1645	29	751	54	479	79	229
5	1480	30	719	55	476	80	224
6	1473	31	714	56	459	81	204
7	1349	32	706	57	454	82	197
8	1336	33	686	58	439	83	191
9	1331	34	685	59	419	84	182
10	1263	35	666	60	418	85	167
11	1251	36	658	61	414	86	161
12	1243	37	648	62	414	87	160
13	1227	38	636	63	397	88	151
14	1158	39	632	64	393	89	147
15	1154	40	625	65	383	90	146
16	1120	41	618	66	375	91	145
17	1115	42	607	67	369	92	135
18	1090	43	597	68	360	93	134
19	1017	44	578	69	359	94	115
20	994	45	541	70	355	95	91
21	915	46	539	71	325	96	83
22	909	47	509	72	322	97	80
23	898	48	508	73	314	98	75
24	881	49	499	74	294	99	68
25	871	50	498	75	290	100	62

dividing by N'. This notation will be used repeatedly throughout this chapter so make sure that you understand it.

An alternative way of quantifying the typical mass is to use the **median** value. This is obtained by **ranking** the pebbles from the heaviest to the lightest and taking the central value. If we had 5 pebbles, the central one would be the third heaviest pebble (also the third lightest). Similarly, for 99 pebbles the median would be the mass of the fiftieth heaviest pebble. However, for an even number of pebbles there is no central pebble. For example, for 4 pebbles the second and third are equally close to being central. In such cases the procedure is to average the two most central values. For 100 pebbles we must average the mass of the fiftieth and fifty-first pebble. Ranking of the beach sample in Table 7.1 results in Table 7.2. Pebbles 50 and 51 are 498 and 496 g respectively. Thus the median mass is the average of 498 and 496, i.e. 497 g. Note that this is quite different from the average mass.

So much for statistics which indicate the typical mass. What about other aspects of the distribution of pebble masses? For example, the pebbles might all have very similar masses or the masses might be very widely dispersed. What is needed is a measure of **dispersion**. A very simple way to indicate this would be to give the **range** of values, i.e. the lowest and highest masses in the sample. In the case of Table 7.1 (or better, Table 7.2) the range is from 62 to 1834 g. However, the heaviest and lightest pebbles might be very untypical. It would be better to use a measure of the spread of masses which is determined by all of the pebbles in the sample rather than a small minority. One such measure is the **mean square deviation from the mean**. This is also called the **variance**. For the total population of pebbles this is denoted by σ^2 and is defined as

$$\sigma^2 = \overline{(\text{Mass} - \text{Average mass})^2} \tag{7.3}$$

where the bar over the expression indicates that the average value of this quantity should be calculated. In other words, we first find the average pebble mass and then calculate the difference between this and the mass of each individual pebble. The result is then squared which gives the square deviation from the mean. Finally, the average value of this for all the pebbles is found. The deviation of each pebble from the mean is squared since some of the deviations are negative (mass less than average) and some are positive (mass higher than average) leading to an average deviation of zero. Squaring all the deviations from the mean ensures that the average of a series of positive numbers is found which will, of course, also be a positive number. Notice that if the masses are all very similar then they will all be very close to the mean leading to a small value for the variance. If, on the other hand, the masses differ widely from one another then some of these masses will be a long way from the average value and the variance will be much larger. The standard deviation, σ, which is simply the square root of equation (7.3), could also be used to indicate the range of values in the population.

However, it would be better to have a measure of distribution width based upon a sample rather than the entire population. An obvious candidate would be the **sample variance**, s^2, i.e. apply equation (7.3) to a sample rather than the population. In terms of the notation introduced earlier this gives

$$s^2 = \frac{1}{N}\left(\sum_{i=1}^{N} (w_i - \bar{w})^2\right) \tag{7.4}$$

where $\bar{w}$ is the average mass defined by equation (7.2). There is a slightly easier method for calculating s^2 since equation (7.4) can be rearranged to give

$$s^2 = \overline{w^2} - (\bar{w})^2 \tag{7.5}$$

i.e. the mean of the squared masses minus the square of the mean mass (proof of this is given as an exercise at the end of the chapter). Using the figures from Table 7.1, the mean square mass is 560 117 g^2 whilst the square of the mean is $610^2 = 372\,100\,g^2$. Thus, the sample variance is $560\,117 - 372\,100 = 188\,017\,g^2$.

As a measure of the width of the distribution this number has several disadvantages. The first problem is that $\bar{w}$ itself has been estimated from the sample. In fact, the estimate of $\bar{w}$ obtained from equation (7.2) has the property that it is the value which minimizes the sample variance. If another value is used in equation (7.4) in place of $\bar{w}$ a larger value for s^2 is always obtained. Hence, if the true population mean were substituted into equation (7.4) instead of its estimate, $\bar{w}$, a larger value for s^2 would result. Equation (7.4) is therefore **biased** towards a smaller value than the true one. To counteract this effect the **unbiased estimate**, $\hat{s}^2$ is used instead, where

$$\hat{s}^2 = [N/(N-1)]s^2. \tag{7.6}$$

Note that for large sample sizes this increases the variance very slightly whereas for smaller sample sizes this variance estimate is significantly larger than s^2. A formal proof that $\hat{s}^2$ is a better estimate of the population variance than s^2 is beyond the scope of this chapter but you should be able to see that it has the effect of altering s^2 in the right direction (i.e. it increases it by an amount which depends upon the sample size). Table 7.1 has a value for N of 100; the figure calculated above for s^2 then produces an estimate of the population variance of $\hat{s}^2 = 100 \times 188\,017/99 = 189\,916\,g^2$.

Another problem with using variance to measure distribution width is that the final number (in this case $189\,916\,g^2$) is not very understandable. What does this result actually mean? Perhaps the simplest way to look at this is to use the variance just for comparison purposes. Take two samples of 100 pebbles from two different beaches. If the first beach has a larger variance for the masses than the second, the pebbles on the second beach tend to have more similar masses to each other than those from the first beach.

The 'interpretability' of variance is further improved by taking its square root. This gives an estimate, $\hat{s}$, of the population standard deviation σ. In the case of our data this yields an answer of $\sqrt{189\,916} = 436\,g$. For reasons that fall outside the scope of this chapter, this result implies that around 68% of all pebble weights should fall within 436 g of the mean value (610 g). Thus, 68% of all weights should fall between 174 and 1046 g. In fact, out of the 100 measurements in Table 7.1, 66 fall in this range which is of course 66% of the total. Thus the theoretical prediction that 68% of the pebble weights should fall in this range is not at all bad! Standard deviation is therefore a very simple way of describing the range of values in your data. A large standard deviation implies a wide spread of values whilst a small standard deviation implies a small spread.

Standard deviation is probably the most common measure of variability that you will encounter. If, in a particular case, no indication is given as to the measure used, you can reasonably assume that standard deviation is implied. For example, a geochemical analysis might quote the amount of lead present in a mineral as $5.2 \pm 0.5\%$. In this case, the statement implies that the average from a large number of measurements is 5.2% and the variation between measurements is described by a standard deviation of 0.5%. Note that the reason for variation from specimen to specimen in an example like this could be due to measurement errors or to real variability in the mineral composition.

Question 7.3 Using the first 10 values from Table 7.1, calculate:

(i) the sample mean,
(ii) the sample median,
(iii) the sample variance.

Also estimate:

(iv) the population variance and standard deviation.

Compare these results to those obtained above from the sample of 100 pebbles.

There are many other parameters and statistics which could be calculated for given populations and samples. However, the most important are undoubtedly the mean and the standard deviation and these are the ones which you should be most familiar with.

7.4 HISTOGRAMS

It is useful to have a method for displaying, for example, the distribution of pebble masses in Table 7.1 graphically. The simplest such method uses the **frequency histogram** which allows the general properties of the distribution to be visualized. To construct such a plot we must first count the number of occurrences of pebble masses within specified ranges. For example, there are

Table 7.3 Number of pebbles in Table 7.1 which fall into 200 g-wide classes

Range (g)	*Number of occurrences*
1–200	19
201–400	19
401–600	20
601–800	16
801–1000	7
1001–1200	6
1201–1400	7
1401–1600	2
1601–1800	2
1801–2000	2

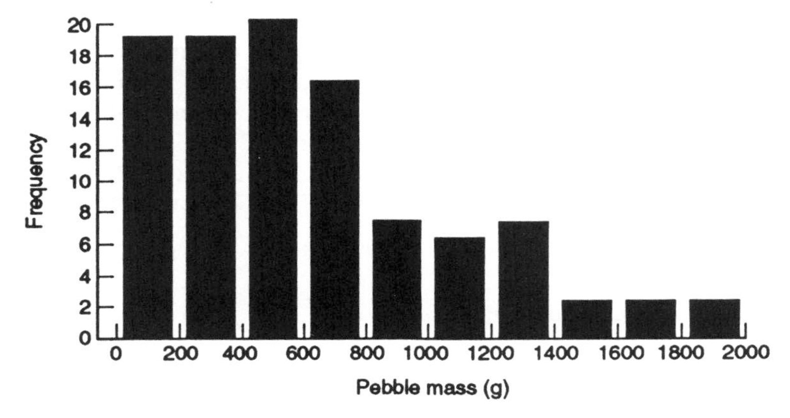

Figure 7.1 The frequency histogram resulting from counting the number of pebbles from Table 7.1 in each 200 g-wide class.

eight pebbles in Table 7.1 with masses between 201 and 300 g. Table 7.3 lists the number of pebbles with masses in 200 g-wide **classes** from 1 to 2000 g. Such a table is called a **frequency distribution**. If these figures are plotted as a bar chart the result is a frequency histogram (Figure 7.1). From this we can instantly see that the masses most commonly fall between 400 and 600 g and are heavily skewed towards the low end.

It was pointed out in Chapter 5 that, if your data are a function of a cyclic variable such as direction or longitude, they are best represented by a polar plot. The same is true for histograms. For example, cross-bedding and ripple marks in sandstones can be used to indicate the **palaeocurrent direction**, i.e. the direction of transport of ancient rivers or submarine currents. Such data will usually have considerable scatter due to uncertainties in measurement and the effect of local topographic features. Thus, if the general direction of

Table 7.4 Number of palaeocurrent measurements as a function of direction

Direction range (°E of N)	*Number of measurements*
1–30	43
31–60	23
61–90	10
91–120	11
121–150	14
151–180	20
181–210	10
211–240	4
241–270	15
271–300	20
301–330	40
331–360	36

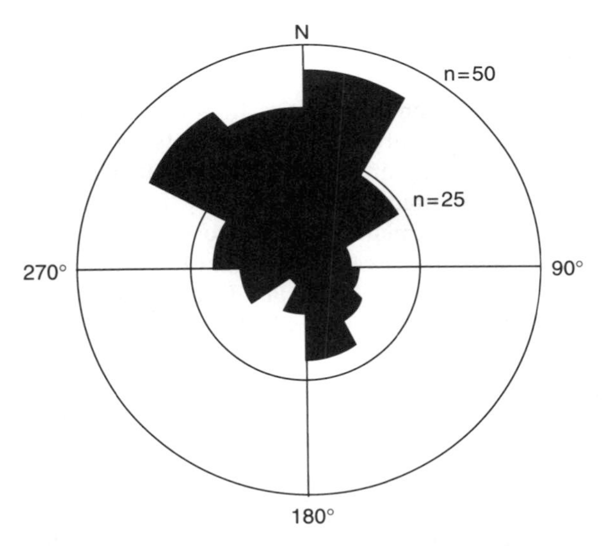

Figure 7.2 A 'rose diagram' in which frequency as a function of direction is represented by distance from the centre.

transport is required, the best course of action is to collect a large number of measurements and then to plot these on a histogram. Table 7.4 lists the frequency data from such a series of measurements.

Now, an obvious way to plot these data is as a histogram on polar graph paper. In other words, plot the frequency as a function of direction such that direction is represented by angle around the plot and the frequency is proportional to distance from the plot centre. This yields a **rose diagram** (Figure 7.2) from which it is very clear that the main current direction was roughly NNW.

Table 7.5 Estimates of the probability of a pebble picked at random being in a specific range. Data modified from Table 7.3

Range (g)	*Probability*
1–200	0.19
201–400	0.19
401–600	0.20
601–800	0.16
801–1000	0.07
1001–1200	0.06
1201–1400	0.07
1401–1600	0.02
1601–1800	0.02
1801–2000	0.02

7.5 PROBABILITY

Probability is a central concept in statistics. In essence, the idea is very simple. If I perform a very large number of measurements on field data or experimental data then I can determine how often a particular result is obtained. This will then allow me to predict the probability that a particular result will occur in any future measurement. Thus, if I toss a dice 1000 times and the number two occurs 400 times, I can predict that the probability is 0.4 of two being the result of my next throw of the dice. I can also conclude that the dice is probably loaded! Note that an event which has a probability of 1 is certain to occur whilst an event whose probability is zero will never occur.

Similarly, for the data shown in Table 7.3 and Figure 7.1, the most probable weight range (401–600 g) occurs in 20 cases out of 100, i.e. 20% of the time. In other words, an estimate of the probability of a pebble, picked up at random from the beach, being in the range 401–600 g is 0.2. Repeating this procedure for the entire table leads to Table 7.5 which is a **probability distribution**. The results can then be plotted in a new histogram (Figure 7.3). Note that the shape of this is identical to Figure 7.1 except that the vertical scale has been shrunk by a factor of 100 (i.e. divided by the size of the sample).

This probability distribution can be compared to various theoretical distributions. The most important of these is the **normal distribution** otherwise known as the **Gaussian distribution**. This has the form

$$P(x) = \frac{\exp[-(x-\bar{x})^2/2\sigma^2]}{(2\pi)^{1/2}\sigma} \qquad (7.7)$$

where $P(x)$ is called the **relative probability** of obtaining the value x, x is the average of all x-values and σ is the standard deviation of the distribution. A graph of this function is shown in Figure 7.4 for the case of mean equal to 5.0

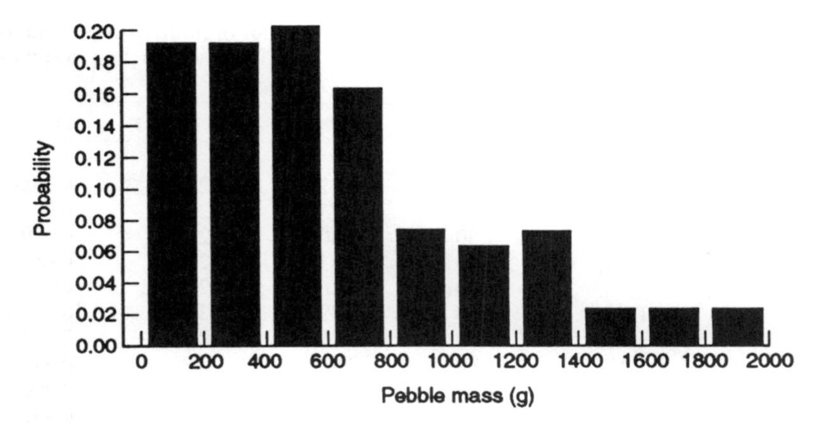

Figure 7.3 Probability distribution histogram for the pebble masses from Table 7.1. Note that the form of this is identical to Figure 7.1 except that the vertical scale has been shrunk by a factor of 100 (equal to the total number of pebbles in the sample).

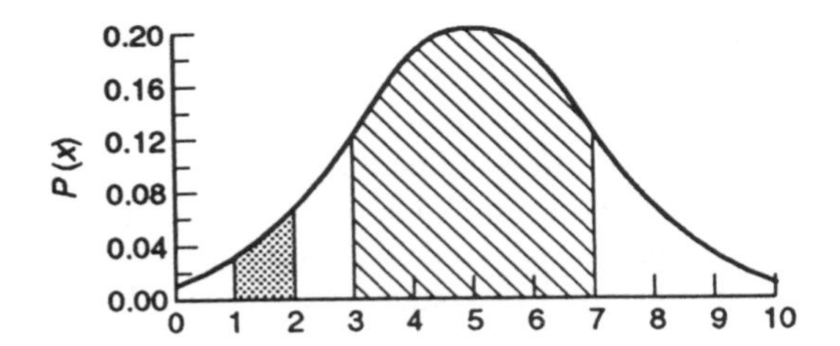

Figure 7.4 Gaussian probability distribution for a mean of 5.0 and standard deviation of 2.0. The diagonally shaded region is the area under the graph within one standard deviation of the mean. The area of the stippled region gives the probability of a result lying between 1.0 and 2.0. The total area under the graph is exactly 1.0.

and standard deviation equal to 2.0. Note that the maximum probability occurs at the mean value, that the curve is symmetrical about the mean and that a large fraction of the area under the graph occurs between $\bar{x} - \sigma$ and $\bar{x} + \sigma$ (i.e. the diagonally shaded area).

The probability of obtaining values within a specified range is governed by two things. Firstly, the higher the graph is within that range, the higher the probability is. Thus, given the graph shown in Figure 7.4, you are more likely to obtain a value between, say, 5 and 6 (where the graph is high) than you are between 1 and 2. Secondly, the probability of obtaining a value within a specified range increases as the width of the range increases. Thus, you are more likely to obtain a value between 5 and 7 than between 5 and 6 because more possibilities are included in the second case. In fact, the relative probability distribution is defined such that the probability of obtaining a

Table 7.6 The area under the Gaussian curve as a function of multiples of standard deviation (s.d.) from the mean. For example, the area lying within two standard deviations of the mean is 0.954

No. of s.d.s	*Area*	*No. of s.d.s*	*Area*	*No. of s.d.s*	*Area*
0.00	0.000				
0.10	0.080	1.10	0.729	2.10	0.964
0.20	0.159	1.20	0.770	2.20	0.972
0.30	0.236	1.30	0.806	2.30	0.979
0.40	0.311	1.40	0.838	2.40	0.984
0.50	0.383	1.50	0.866	2.50	0.988
0.60	0.451	1.60	0.890	2.60	0.991
0.70	0.516	1.70	0.911	2.70	0.993
0.80	0.576	1.80	0.928	2.80	0.995
0.90	0.632	1.90	0.943	2.90	0.996
1.00	0.683	2.00	0.954	3.00	0.997

value within a given range is given by the area under the graph over that range. For example, the probability of obtaining a value between 1.0 and 2.0 is given by the area of the stippled region in Figure 7.4 whilst the (much greater) probability of obtaining a value between 3 and 7 is given by the area of the diagonally shaded region. Given this definition, the total area under the graph is 1.0 since the probability of obtaining some value is 1.0. This is indeed the case for the Gaussian distribution defined in equation (7.7).

So, how can such areas be found? The simplest method is to use a table showing area under the curve as a function of multiples of standard deviation from the mean (e.g. Table 7.6). From such a table it can be seen that the area of the diagonally shaded region in Figure 7.4 is 0.683. Similarly, the area under the curve within two standard deviations (i.e. between 1.0 and 9.0 in the example shown in Figure 7.4) is 0.954.

The table can also be used to find areas such as the stippled zone in Figure 7.4. To do this, you should note that 1.0 is two standard deviations from the mean whilst 2.0 is 1.5 standard deviations from the mean. From Table 7.6 it can be seen that the area within 1.5σ is 0.866 whilst the area within 2σ is 0.954. Thus, the area between 1.5σ and 2.0σ is $0.954 - 0.866 = 0.088$. However, there are two such zones, one between 1.0 and 2.0 and the other between 8.0 and 9.0 (Figure 7.5). The area between 1.5σ and 2.0σ is shared equally between these two regions and thus the area of the zone between 1.0 and 2.0 is half of 0.088 (i.e. 0.044). Thus, if a process has a probability distribution like that shown in Figure 7.4, the probability of obtaining a result between 1.0 and 2.0 is 0.044.

Thus, if a population variable has a Gaussian probability distribution, the probability of obtaining results within specified ranges can be calculated. However, the Gaussian distribution function is an idealized model. Real populations are often approximately Gaussian but never exactly so. How good is the Gaussian model at predicting the probabilities shown in Table 7.5

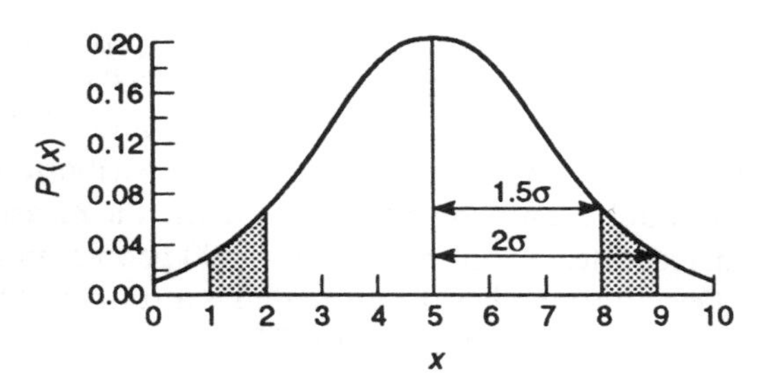

Figure 7.5 The area under the Gaussian curve between 1.5 standard deviations from the mean and 2.0 standard deviations from the mean. Note that there are two such zones and the area of each is therefore half that given using Table 7.6.

Table 7.7 Comparison of the probabilities estimated in Table 7.5 with the probabilities estimated assuming a Gaussian distribution

Range (g)	*Measured probability*	*Range (multiples of σ)*	*Gaussian probability*
1–200	0.19	– 1.4 to – 0.94	0.090
201–400	0.19	– 0.94 to – 0.48	0.14
401–600	0.20	– 0.48 to – 0.02	0.18
601–800	0.16	– 0.02 to 0.44	0.18
801–1000	0.07	0.44 to 0.90	0.15
1001–1200	0.06	0.90 to 1.36	0.098
1201–1400	0.07	1.36 to 1.81	0.053
1401–1600	0.02	1.81 to 2.27	0.026
1601–1800	0.02	2.27 to 2.73	0.0059
1801–2000	0.02	2.73 to 3.19	0.0027

for the pebble weights on our beach? Now, the mean and standard deviation for the pebble weights have been estimated to be 610 and 436 g respectively. Using these values, Table 7.6 can be used to predict the probabilities in various ranges of weights. The results are shown in Table 7.7 together with the measured probabilities shown in Table 7.5 (in fact I have used a slightly more detailed table than 7.6).

Question 7.4 The range from 801 to 1000 g starts 0.44 standard deviations above the mean and ends 0.90 standard deviations above the mean. Using Table 7.6 and these values, estimate the Gaussian probability of a pebble weight lying in this range. N.B. To get the area under the curve within 0.44 standard deviations (call it $P_{0.44}$) assume that it is given by the expression

$$P_{0.44} \approx 0.6P_{0.4} + 0.4P_{0.5}$$

i.e. an average of the probabilities corresponding to 0.4σ and 0.5σ weighted towards the 0.4σ probability.

These results are perhaps best described as patchy. The probabilities measured and predicted are quite close in some cases (e.g. 601–800 g and 1401–1600 g) but in other ranges are not at all close (e.g. 1–200 g and 1801–2000 g). The mismatches have two possible causes. Firstly, the distribution may not be Gaussian at all. Secondly, sampling effects may cause the measured probabilities to be very inaccurate, i.e. other samples of 100 pebbles might give quite different results. Deciding whether the mismatch is a statistical fluctuation or a real difference is briefly covered in section 7.7.

7.6 'BEST FIT' STRAIGHT LINES

So much for statistical analysis of a single variable. What about analysis of two related variables? As discussed in Chapter 2, it is very common for graphs of the relationship between pairs of geological variables to be well approximated by straight lines. However, the fit is never perfect. Thus, a straight line must be found which passes as close to all the data points as possible. The problem of how to find this line is ideally suited to a statistical treatment.

First, we have to define what we mean by a 'best fit' straight line. The usual definition is that the mean square difference between the data and the straight line should be a minimum. Figure 7.6 illustrates this idea. The graph of y as a function of x consists of eight points and a straight line has been drawn which passes close to all of these. The deviation, Δy, of the last point from the straight line is also shown. The mean square deviation is found by calculating the square of this distance for all of the points and then finding the average. Now, if the line is a poor fit to the data, Δy for many of the points will be large and the average squared value will also be large. A good fit for the straight line will result in a much smaller average. The best fit straight line is defined

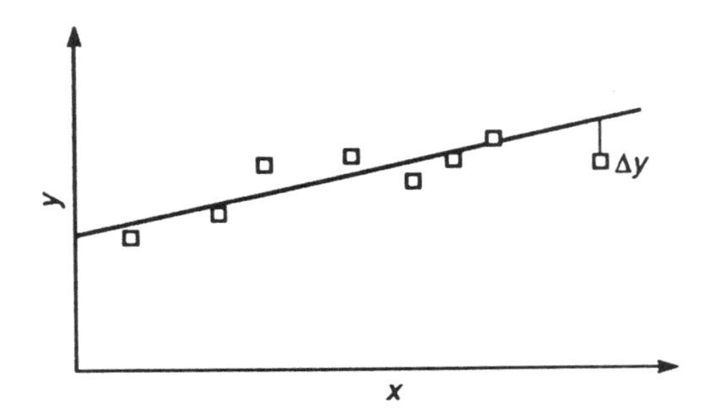

Figure 7.6 A straight line drawn through the x–y data such that it passes close to all of the points. The deviation, Δy, of the line from the last point is also shown. The deviation could be found for each of the other seven points as well. The best fit straight line is that which produces the smallest possible $(\Delta y)^2$ averaged over all of the points.

as that line which results in the smallest possible **mean square deviation**. The process of finding such a line is called **linear regression**.

We now need a method for estimating the gradient, m, and intercept, c, of this straight line. A formal proof is, again, beyond the scope of this chapter but the result is that the best gradient is given by

$$m = \frac{n\sum xy - \sum x \sum y}{n\sum x^2 - (\sum x)^2}. \tag{7.8}$$

In this expression all summations are evaluated for n measurement pairs (x, y). The first summation, $\sum xy$, is simply the sum of all products x_1y_1, x_2y_2 up to x_ny_n. Similarly, $\sum x$ is the sum of all the x-values, $\sum y$ is the sum of all the y-values and $\sum x^2$ is the sum of all the x-values squared. The best estimate for the intercept then follows directly from the fact that the best fit line passes through the point $(\bar{x}, \bar{y})$, i.e. the point defined by the average x-value and the average y-value. Thus

$$\bar{y} = m\bar{x} + c$$

giving

$$c = \bar{y} - m\bar{x}. \tag{7.9}$$

Table 7.8 shows the age versus depth data used in question 2.11 but with the first point excluded. Now, given that the remaining data when plotted seemed to fit a reasonably good straight line, what is the best fit line through these points? First, we should construct all the summations given in equation (7.8). In this example, we need to plot age as a function of depth and therefore 'Depth' replaces x and 'Age' replaces y. The summations required by equation (7.8) are then

$$\sum \text{depth} = 407 + 545 + 825 + 1158 + 1454 + 2060 + 2263 = 8712\,\text{cm} \tag{7.10}$$

$$\sum \text{age} = 10\,510 + 11\,160 + 11\,730 + 12\,410 + 12\,585 + 13\,445 + 14\,685 = 86\,525\ \text{years} \tag{7.11}$$

Table 7.8 The age versus depth data used at the end of Chapter 2. This time the first data point has been excluded since it did not fall close to a straight line defined by these remaining points

Depth (cm)	*Age (years)*
407.0	10 510
545.0	11 160
825.0	11 730
1158.0	12 410
1454.0	12 585
2060.0	13 445
2263.0	14 685

$$\sum \text{depth} \times \text{age} = (407 \times 10\,510) + (545 \times 11\,160) + (825 \times 11\,730) + (1158 \times 12\,410) + (1454 \times 12\,585) + (2060 \times 13\,445) + (2263 \times 14\,685)$$
$$= 4\,277\,570 + 6\,082\,200 + 9\,677\,250 + 14\,370\,780 + 18\,298\,590 + 27\,696\,700 + 33\,232\,155$$
$$= 113\,635\,245 \text{ cm y} \tag{7.12}$$

$$\sum (\text{depth}^2) = 407^2 + 545^2 + 825^2 + 1158^2 + 1454^2 + 2060^2 + 2263^2$$
$$= 13\,963\,148 \text{ cm}^2. \tag{7.13}$$

From equations (7.10) and (7.11), together with the fact that there are seven measurements, the mean depth and mean age are

$$\overline{\text{Depth}} = 8712/7 = 1245 \text{ cm} \tag{7.14}$$

and

$$\overline{\text{Age}} = 86\,525/7 = 12\,361 \text{ years}. \tag{7.15}$$

Thus, substituting results (7.10)–(7.13) into equation (7.8) gives

$$m = \frac{7 \times 113\,635\,245 - 8712 \times 86\,525}{7 \times 13\,963\,148 - 8712 \times 8712}$$
$$= 1.91 \text{ y cm}^{-1}. \tag{7.16}$$

Substituting this into equation (7.9) gives the best estimate of the intercept as

$$c = \overline{\text{Age}} - (m \times \overline{\text{Depth}})$$
$$= 12\,361 - (1.91 \times 1245)$$
$$= 9983 \text{ years}. \tag{7.17}$$

The data from Table 7.8 are plotted in Figure 7.7 together with a straight line of gradient 1.91 y cm^{-1} and intercept 9983 years. As you can see, the fit is

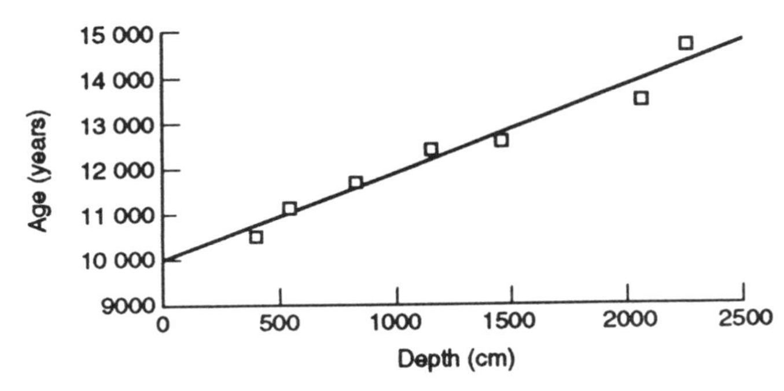

Figure 7.7 The data from Table 7.3 plotted for comparison with a best fit straight line calculated to have a gradient of 1.91 y cm^{-1} and an intercept of 9983 years.

remarkably good. Note that, before starting this exercise, I deliberately excluded a point that did not fit the general linear trend. This means that the resultant line is not appropriate at ages close to this point (i.e. for sediments less than about 10 000 years old). However, it was important to exclude this point since fitting a straight line through points which do not have a simple linear trend would be a meaningless exercise.

Question 7.5 Calculate the best fit straight line through the data in Table 2.2.

7.7 HYPOTHESIS TESTING

This section is about methods for determining the probability that a statement is correct. Hypothesis testing is one of the most important parts of statistics to the geologist. However, a proper treatment of this subject requires an entire textbook to itself. In this section I attempt to give an introduction to the underlying concepts of this branch of statistics. The examples I have chosen are designed to illustrate these ideas rather than as realistic examples of the problems actually tackled using hypothesis testing.

A medical scientist attempting to establish whether there is a link between diet and a particular disease is a good example of hypothesis testing. This would be done by attempting to demonstrate that there is no link between diet and the disease. This is called a **null hypothesis**. A null hypothesis is usually a 'hypothesis of no difference' which, in this case, is that 'diet makes no difference to the incidence of the disease'. Other examples of null hypotheses will be given later.

The next step is to attempt to reject the null hypothesis. This is done by attempting to demonstrate that any apparent link shown by the data could not, with reasonable probability, have occurred purely as a result of random fluctuations. To do this, we simply calculate the frequency with which random fluctuations in the data would produce an apparent link at least as strong as that observed. If the probability of an apparent link being a random effect falls below a predefined **significance level**, the null hypothesis is rejected. In the example used here, if the probability of the observed link occurring by chance is 2% and a 5% significance level was chosen, the null hypothesis would be rejected. Thus, the link would be regarded as accepted.

The concepts of 'null hypothesis' and 'significance level' are very important so it is advisable to make sure you understand them.

Question 7.6 A palaeontologist is attempting to determine whether a particular fossil specimen is an *Orthoceras* by comparing its length to those of known *Orthoceras* specimens.

(i) What would be a good null hypothesis for this problem?

(ii) The specimen was 1.5 times as long as an average *Orthoceras* and the probability of this occurring for a true *Orthoceras* was given as 10%. Should the specimen be accepted as an *Orthoceras* if a 5% significance level and your null hypothesis are being used?

To illustrate hypothesis testing, I will use an example based upon the beach pebbles discussed in section 7.2. The sample of 100 pebbles was found to have a mean mass of 610 g based upon a sample of 100 pebbles. Suppose that there are some theoretical grounds for expecting the beach to have pebbles with a mean mass of 550 g and a standard deviation 400 g (perhaps based upon a study of tides and currents in the area). Thus, the observed mean is 60 g heavier than the value predicted by the theory. Now, statistics vary slightly from sample to sample so is the observed mean consistent with the predicted value or is the theory wrong?

The null hypothesis, in this example, is the statement that there is no difference between the population from which the sample of 100 pebbles was taken and a population with a mean of 550 g and a standard deviation of 400 g. We need to calculate the probability that a sample of 100 pebbles from such a population would have a sample mean deviating 60 g or more from the population mean. If this probability is low, the null hypothesis will be rejected as unlikely and we should conclude that the theory is wrong.

So, how much do sample means tend to differ from population means? Imagine collecting a large number of samples each consisting of n pebbles taken from the same population. The means of these samples will be slightly different and each will deviate from the true population mean. Populations with a large standard deviation will give rise to a wider scatter in the sample means than populations with a small standard deviation. In addition, the sample averages will tend to be closer to each other for larger sample sizes. The simplest way to specify the scatter in resultant mean weights is to give the standard deviation of the sample means. This figure is called the **standard error**, s_e, and is given by

$$s_e = \sigma/\sqrt{n} \tag{7.18}$$

where σ is the population standard deviation and n is the size of the sample (proof of this statement is well beyond the scope of this brief introduction). Note that this definition does indeed have the expected dependency on standard deviation and sample size. Thus, in the case of the problem we are considering, the standard error of the theoretical population is

$$s_e = 400/10 = 40\text{ g} \tag{7.19}$$

and therefore, if the pebble weight theory is correct, 68% of all random samples of 100 pebbles should have a mean within 40 g of 550 g (cf. section 7.3 on standard deviation where it was stated that roughly 68% of all measurements should fall within 1 standard deviation of the mean).

> Question 7.7 The pebble masses from Table 7.1 were drawn from a population whose mean and standard deviation (as well we could estimate them) were 610 and 436 g respectively. Assuming these figures are correct, what would you expect the standard error for a sample of 10 pebbles to be? Calculate the mean mass for any 4 samples of 10 pebbles drawn from Table 7.1. Hence, determine the deviation of these means from 610 g. How do the size of these deviations compare to the standard error (i.e. are they much bigger, much smaller or about the same size)?

In other words, 32% of all samples have means more than 40 g away from 550 g. Clearly, the deviation we actually have (observed mean − predicted mean = 60 g) is larger than the standard error and we can therefore conclude that considerably fewer than 32% of all samples would deviate by 60 g or more. In other words, the null hypothesis can be rejected at the 32% significance level. However, a 32% significance level is considerably higher than that normally used. It would be a rash gambler who concluded from this that the theory of pebble weights is wrong. So, what sort of significance level should be set?

When testing a hypothesis, there are two types of error that can be made. Firstly, we might accept as true a hypothesis which is actually false. Secondly, we might reject a hypothesis which is actually true. The best level of significance to set is determined by the relative importance of these two types of error. If it is very important that we do not accept a false null hypothesis then the significance level should be set high. This will ensure that the null hypothesis is relatively easily rejected. On the other hand, it might be more important that we do not reject a true null hypothesis. In this case the significance level should be small so that it relatively easy to accept the hypothesis. These ideas should be made a little clearer by Figure 7.8 which

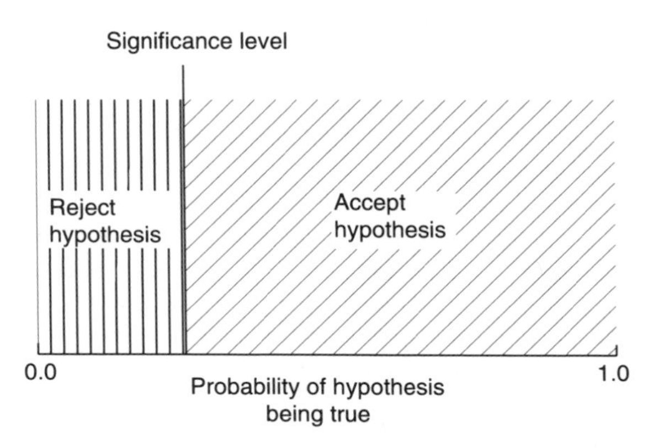

Figure 7.8 The relationship between hypothesis probability, significance level and whether or not the hypothesis is accepted.

Table 7.9 Multiple of standard error as a function of level of significance

Significance (%)	*Number of standard errors*
20	1.29
10	1.64
5	1.96
1	2.58
0.1	3.29

illustrates the relationship between probability, significance level and whether a hypothesis is accepted. The exact level to set is a question of judgement; there are no concrete mathematical rules which can be applied. However, significance levels of 1 or 5% are, perhaps, the most commonly used values. This implies that we are normally making it relatively easy to accept the null hypothesis. Thus, it is very important that a null hypothesis is chosen intelligently!

So, is the pebble weight theory acceptable at the 1 or 5% significance levels? Table 7.9 gives the multiple of standard error necessary to reach various levels of significance. Thus, for example, 20% of all sample means are beyond 1.29σ of the population mean. Our pebble sample deviates from the predicted mean by 1.5 standard errors ($1.5 \times 40 = 60$ g). Thus, the null hypothesis should be accepted at both the 1 and 5% levels. At this stage we would, provisionally, accept the pebble weight theory as consistent with the data.

Question 7.8 Would you also accept the pebble weight theory at the 0.1% significance level?

7.8 MORE ADVANCED HYPOTHESIS TESTING

The above example is actually a very simple case of hypothesis testing. In other cases we might wish to test whether a particular fossil is an *Orthoceras*, whether a sedimentary sequence is random or whether the distribution of pebble masses has a particular shape. In each of these cases a slightly different type of test is necessary. Statistics textbooks are full of many such tests and a large part of the skill in using statistical methods consists of deciding what you wish to prove, designing appropriate fieldwork or experiments and selecting the best statistical test. However, in nearly all cases the statistical procedure can be broken down as follows:

1. The first step in devising a test is to set up a null hypothesis that there is no difference between the population being investigated and some hypothetical population. Thus, for fossil identification, the null hypothesis is that

the specimen is an *Orthoceras*. For the sedimentary sequence case, the null hypothesis is that the sequence is random. For the pebble mass case, the null hypothesis might be that the masses have a Gaussian distribution. Note that, associated with the null hypothesis, there is a theoretical population for which the null hypothesis is true. In the fossil identification case this is the population of all *Orthoceras* specimens. In the sedimentary sequence case, this is the population of all random sedimentary sequences. In the pebble distribution case, this is all possible pebbles drawn from a Gaussian distributed population.

2. Next, a level of significance is chosen. It is good practice to decide this in advance. You should never adjust the level of significance to aid acceptance of a pet theory!
3. Having set up a null hypothesis and a significance level we must now choose a statistical test. This determines the probability that the observed deviations from the null hypothesis could have occurred by chance if the sample really came from the theoretical population. If this probability is less than the significance level we have set then the null hypothesis is rejected and the sample is assumed to be from a different population. If this probability is higher than the significance level we set then the null hypothesis is accepted and the sample is assumed to be from the theoretical population.

FURTHER QUESTIONS

7.9 Measured dip directions for a particular bed at 20 different locations in a field area are given in Table 7.10.

(i) Use these data to calculate the frequency distribution and estimate the probability distribution using 45° wide classes.

Table 7.10 Dip directions for one bed at different locations within a field area

Location No.	*Orientation (°E of N)*	*Location No.*	*Orientation (°E of N)*
1	27	11	14
2	63	12	355
3	10	13	300
4	87	14	96
5	103	15	276
6	256	16	190
7	200	17	191
8	23	18	23
9	17	19	10
10	25	20	5

(ii) Plot the frequency distribution on a suitable type of histogram.
(iii) Is there an overall trend for these data?

7.10 In section 7.3 it was stated that

$$s^2 = \frac{1}{N}\left(\sum_{i=1}^{N} (w_i - \bar{w})^2\right) \quad (7.4)$$

can be rearranged to give

$$s^2 = \overline{w^2} - (\bar{w})^2. \quad (7.5)$$

Verify this by:

(i) Multiply out $(w_i - \bar{w})^2$ in equation (7.4).
(ii) Split the result into three separate summations using the following relationship:

$$\sum(a + b + c + \cdots) = \sum a + \sum b + \sum c + \cdots$$

Table 7.11 The percentage weight of calcium carbonate and organic carbon at various depths in the top 100 cm of core RAMA 44PC

$CaCO_3$ (%)	*Organic C (%)*
6.10	0.35
5.30	0.27
5.30	0.28
6.70	0.35
9.00	0.42
7.20	0.43
3.20	0.22
14.3	0.39
13.4	0.48
15.3	0.68
9.00	0.70
3.40	0.87
6.30	0.86
10.2	0.95
10.5	0.95
13.4	1.23
15.1	1.22
5.70	1.25
1.90	1.05
2.00	0.98

Data from Keigwin, L., Jones, G. and Froelich, P. (1992). A 15 000 year palaoenvironmental record from Meiji Seamount, far north-western Pacific. *Earth and Planetary Science Letters*, **111**, 425–40.

(iii) Simplify each of the resulting summations using the following result:

$$\sum ka = k\sum a \quad \text{where } k \text{ is a constant.}$$

(iv) Finally, use the definition of mean and mean square to simplify further and obtain the required result.

7.11 Skewness has been mentioned several times in this chapter and is a measure of the symmetry of a distribution. One of several possible definitions is

$$\text{Skew} = \left(\sum (w_i - \bar{w})^3\right)/Ns^3 \qquad (7.20)$$

which will equal zero if and only if the distribution is symmetric.

(i) Evaluate this expression for the same 10 pebbles you used in question 7.3.
(ii) Comment on your answer in the light of Figure 7.1.
(iii) I might wish to test whether the population from which the 10 pebbles were taken was really skewed or whether the apparent skew is a statistical fluctuation. What null hypothesis should I use?

7.12 Table 7.11 lists the calcium carbonate and organic carbon weight percentages obtained at various points within the top 100 cm of a core from a seamount in the north-west Pacific Ocean.

(i) Calculate a linear regression for these data.
(ii) Plot the resulting best fit straight line together with a scatter plot of the original data (i.e. plotted as individual points, not joined together).
(iii) How well do you think a linear regression works in this case?

8 Differential calculus

8.1 INTRODUCTION

Calculus, discovered in the seventeenth century by Newton and independently by Leibniz, is where advanced mathematics is usually assumed to begin. Despite this, the mathematical manipulations used in elementary calculus are very simple to apply. The real difficulty with calculus is not in 'How is it done?' but rather in 'How is it used for solving a particular problem?' A further difficulty is the, at first sight, strange new notation that is required. The mathematical expressions used in calculus are quite different from those you will have encountered previously and this is a significant barrier to understanding. In other words, calculus calculations are easy to do once you understand their strange appearance but this knowledge is of limited use until you have a good grasp of how to use mathematics more generally for solving real problems. Hopefully, the preceding chapters will have improved your ability to apply simple mathematics. In this chapter, and the next, I shall introduce you to the techniques and notation of calculus and show a few examples of ways to use these techniques. First, however, I should discuss what calculus is for.

8.2 RATES OF GEOLOGICAL PROCESSES

Geology is a process-centred subject. Virtually all geological observations can, and should, be explained by the interaction of various physical, chemical or biological processes. Increasing burial, for example, is a process which has the effect of compressing, heating and chemically altering a sediment. Biological activity such as burrowing is a process which has the effect of mixing the uppermost metre or so of sediment. Convergence of two plate tectonic plates is a process which can lead to the formation of mountain chains. The point that I wish to make is that all of these processes, and their results, occur at a certain rate. Burial, for example, might occur at a rate of 1 m per 1000 years or so. Biological mixing could lead to, say, 10% of the upper layer being reworked every year. Plate convergence rates are generally several centimetres per year. These are all examples of rates expressed in terms of the amount of change produced in a certain amount of time, e.g. 1 m of extra sediment in 1000 years, 10% mixing in one year or 3 cm of plate convergence in one year.

Compaction of rocks during burial could be expressed in terms of loss of porosity per year (e.g. porosity might reduce by 0.1% per year in a particular case). However, in many cases it is more useful to express compaction in terms of the loss of porosity produced by a given amount of burial. A typical figure might be a 12% loss in porosity produced by 1 km of burial. In this case, the rate of change of porosity is expressed in tems of the amount of change produced in a certain distance.

The key concept here is that we are considering the amount of change in a quantity (e.g. porosity) produced by change in another quantity (e.g. depth of burial). The main difficulty is that, for most processes, this rate of change is not a constant. There might be a loss of 12% porosity as a result of burial by 1 km but further burial down to 2 km might only result in the loss of an additional 5% porosity. In this example, the rate of loss of porosity is faster near the surface than it is at depth. Returning to the near-surface mixing caused by biological activity, this is likely to occur at a faster rate in the summer than it does in the winter so we again have a process whose rate is not a constant.

To avoid this difficulty, rates of change are normally given in terms of their **instantaneous rate**. For example, the rate of loss of porosity at the surface might be 40% per kilometre. This means that the porosity would be reduced by 40% after burial by 1 km if this rate remained constant down to 1 km. In practice, the rate will probably drop off with depth so that the porosity loss after burial by 1 km will be much less than 40%. A more familiar example is the speed of an automobile. If you are travelling at 50 km $hour^{-1}$ this means that you would move 50 km in the next hour if you continued at the same rate. It does not mean that you will necessarily move 50 km in the next hour. You might get stopped by heavy traffic after 5 minutes!

Calculus is a set of mathematical tools which allow us to deal with processes whose rates are not constants. Since this is the case for most real processes, calculus is an essential part of all applied mathematics. In particular, **differential calculus** is the tool used to determine the instantaneous rate at which a given process takes place.

8.3 GRAPHICAL DETERMINATION OF RATES OF CHANGE

Before we go on to use differential calculus, it is instructive to consider simpler alternatives. This will allow me to introduce some important concepts and is also a relatively simple way to introduce the notation needed for calculus.

In Figure 8.1 I use the porosity, ϕ, versus depth, z, example once more. In

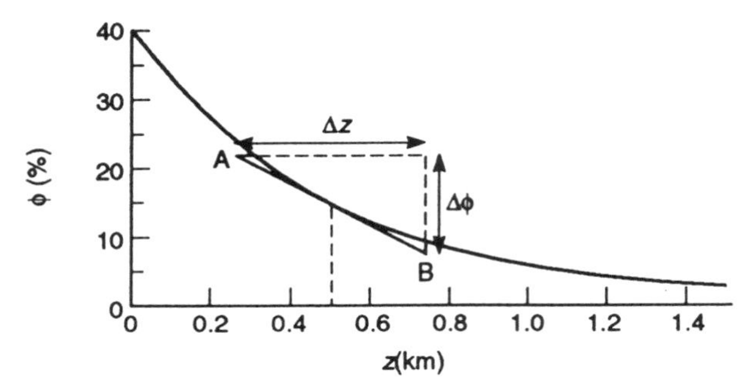

Figure 8.1 A porosity against depth curve. The instantaneous rate of loss of porosity at a depth of 500 m is given by the gradient of the straight line, AB, shown.

the particular example shown, the rate at which the porosity is falling is very fast at the beginning and becomes less fast at greater depths. This is shown on the graph by the fact that the curve is very steep on the left and becomes much flatter towards the right. Thus, the porosity drops by about 7% in the first 100 m of burial but drops by about 0.5% when buried by 100 m starting from a depth of 1300 m. Now, what is the instantaneous rate of porosity loss at a depth of 500 m? To determine this, a **tangent** to the curve at $z = 500$ m is drawn as shown. A tangent is a straight line which touches the curve at one, and only one, point. The rate of change that we wish to find is simply the gradient of this tangent.

Question 8.1 The porosity/depth curve in Figure 8.1 is a smooth graph drawn through the points:

Depth (km)	*Porosity (%)*
0.0	40.0
0.2	26.8
0.4	18.0
0.6	12.0
0.8	8.1
1.0	5.4

Plot these points on a large sheet of graph paper. Draw a smooth curve through the points and measure the gradient of a tangent to the curve at a depth of 500 m.

Now, having determined the gradient at one particular depth, the gradient at several other depths could also be found. I have done this (using a more accurate method you will learn later) for the graph shown in Figure 8.1 and

Table 8.1 The porosity and gradient for various points shown in Figure 8.1

Depth (km)	*Porosity (%)*	*Gradient (% km^{-1})*
0.0	40.0	−80.0
0.1	32.7	−65.5
0.2	26.8	−53.6
0.3	22.0	−43.9
0.4	18.0	−35.9
0.5	14.7	−29.4
0.6	12.0	−24.1
0.7	9.9	−19.7
0.8	8.1	−16.2
0.9	6.6	−13.2
1.0	5.4	−10.8

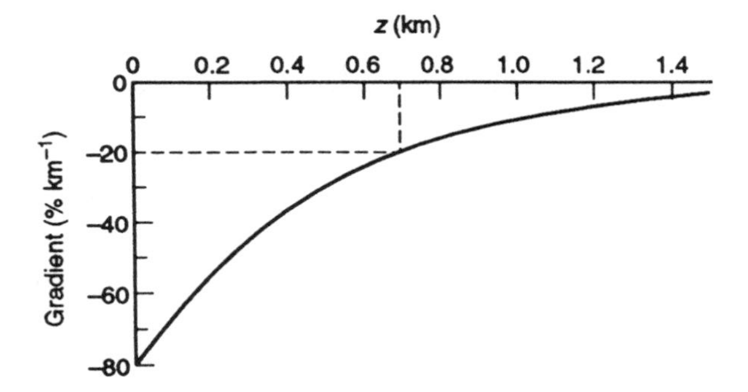

Figure 8.2 Gradient of Figure 8.1 as a function of depth. Hence, at a depth of 700 m the gradient of Figure 8.1 is about −20% km^{-1}.

the result is given in Table 8.1. These gradient values can themselves be plotted on a graph as a function of depth. The result is Figure 8.2. The function in Figure 8.2 is called the **derivative** of the function in Figure 8.1. The important point is that the gradient of a curve is another curve rather than a single number.

8.4 ALGEBRAIC DETERMINATION OF THE DERIVATIVE OF $y = x^2$

The problem with using such graphical approaches is that they are slow and not very accurate. It would be better if there was a more rigorous method for

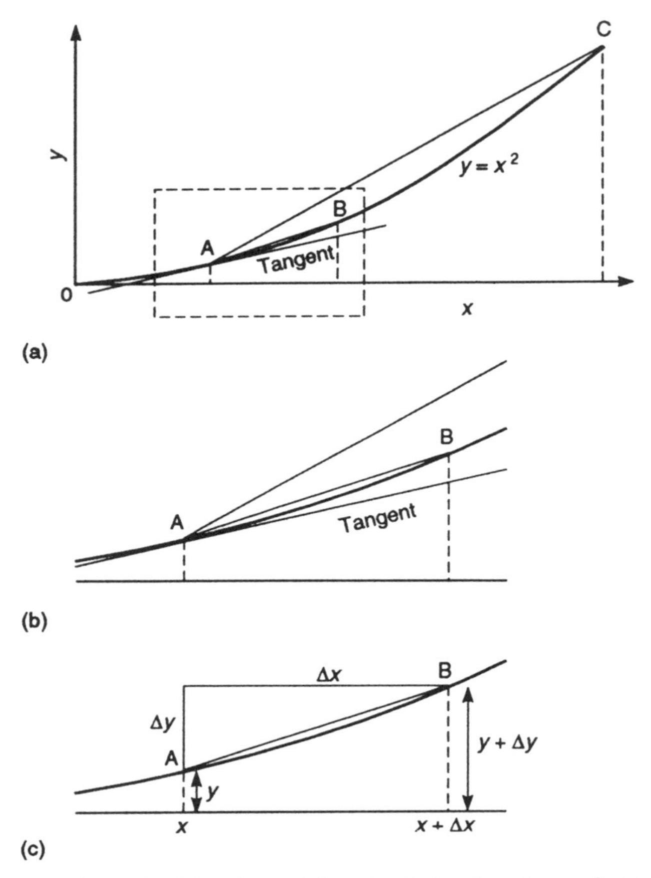

Figure 8.3 An algebraic method for determining the derivative of $y = x^2$. (a) Three points A, B and C plotted on a graph of y against x. (b) Enlargement of the inset from (a). Line AB is closer in gradient to the tangent than the line AC. (c) Point A is at (x, y) and point B is at $(x + \Delta x, y + \Delta y)$. Hence the gradient of line AB is $\Delta y/\Delta x$.

determining the derivative of a given function. One approach would be to use a little algebra to try and calculate the gradient of the tangent. This technique is illustrated in Figure 8.3 for the case of the function

$$y = x^2. \tag{8.1}$$

In Figure 8.3(a), three points, A, B and C, are plotted on the curve. Figure 8.3(b) shows an enlargement around the area of points A and B. In this enlargement it can be seen that the gradient of the line joining A to B is closer to the gradient of the tangent than the gradient of the line AC. The key points here are that we can attempt to estimate the gradient of the tangent by finding the gradient between two points on the curve and that the closer the two points are the better the estimate becomes.

Figure 8.3(c) concentrates on the line joining point A to B. Now, since B is considered to be close to A, we can think of its x-location as being a small amount (Δx) greater than the x-location for A. Similarly, the y-location of B is a small amount (Δy) greater than the y-location of A. A slightly different way to state this is that point A is at the location (x, y) whilst point B is at location $(x + \Delta x, y + \Delta y)$. Using this notation, the gradient of the line AB is just

$$\text{Gradient} = \Delta y / \Delta x. \tag{8.2}$$

Now, since point B is also on the curve given by equation (8.1) it follows that

$$y + \Delta y = (x + \Delta x)^2. \tag{8.3}$$

The right-hand side may then be multiplied out to yield

$$y + \Delta y = x^2 + 2x\Delta x + (\Delta x)^2. \tag{8.4}$$

However, by equation (8.1), we may subtract y from the left-hand side and, simultaneously, remove x^2 from the right-hand side since they are equal. This gives

$$\Delta y = 2x\Delta x + (\Delta x)^2. \tag{8.5}$$

We need the gradient which, by equation (8.2), is given by $\Delta y / \Delta x$. Thus, dividing equation (8.5) by Δx leads to

$$\frac{\Delta y}{\Delta x} = 2x + \Delta x \tag{8.6}$$

which is the gradient of line AB.

However, we wish to have the gradient of the tangent. This is obtained by imagining that point B becomes **infinitesimally** close to point A. This means that there is an unimaginably small distance between A and B but they are not quite touching. Under these conditions Δx is said to **tend** towards zero and the gradient of AB tends to the gradient of the tangent at A. As Δx tends towards zero, equation (8.6) becomes

$$\frac{dy}{dx} = 2x \tag{8.7}$$

in which the symbol dy/dx is used to denote the gradient of the tangent and Δx has disappeared from the right-hand side since it has gone to zero.

Now that we have a formula for the gradient we can use it to determine the slope at any point. For example, at $x = 10$

$$y = x^2 = 10^2 = 100 \tag{8.8}$$

by equation (8.1) and

$$\begin{aligned} \text{Gradient} &= \mathrm{d}y/\mathrm{d}x \\ &= 2x \\ &= 2 \times 10 = 20 \end{aligned} \tag{8.9}$$

by equation (8.7).

Question 8.2 Using the above equation, determine the slope of $y = x^2$ at the points $x = 2$ and $x = 1000$.

Question 8.3 A graph of $y = x^3$ has a gradient given by $\mathrm{d}y/\mathrm{d}x = 3x^2$. Demonstrate this by following through similar steps to those above.

The main point to understand from the above derivation is that calculus obtains its results by considering what happens in the case of infinitesimally small differences between quantities. This is why it is called **differential calculus** and why the above process is called **differentiation**. If the porosity, ϕ, is given as a function of depth, z, the gradient of this function will be given by the function $\mathrm{d}\phi/\mathrm{d}z$ which results from differentiating the function ϕ.

8.5 STANDARD FORMS

In practice it is never necessary to go through derivations, such as those in section 8.4, to obtain the derivative of a given function. This is because mathematicians have already done this for the vast majority of cases that geologists will ever use. The results are summarized in tables of derivatives and it is only necessary to go and look these up. The most common examples of these **standard forms** are given in Table 8.2.

Let us go through some of these just to see if they make sense. Firstly we have that if

$$y = x^n \tag{8.10}$$

Table 8.2 The most commonly encountered standard forms of derivatives

y	dy/dx
x^n	nx^{n-1}
$\sin(x)$	$\cos(x)$
$\cos(x)$	$-\sin(x)$
$\tan(x)$	$1/\cos^2(x)$
e^x	e^x
$\ln(x)$	$1/x$

then

$$\frac{dy}{dx} = nx^{n-1}. \tag{8.11}$$

The simplest way to understand this is to use the example followed in section 8.4. In that section I considered the case of

$$y = x^2. \tag{8.1}$$

This is a particular example of the more general case given by equations (8.10) and (8.11). Comparing equation (8.1) to equation (8.10) you should be able to see that, if $n = 2$, these two are the same. Now, if you replace n by 2 in equation (8.11) you should get

$$\frac{dy}{dx} = 2x^1$$

$$= 2x \tag{8.12}$$

which is the same as the result given by equation (8.7).

Question 8.4 Using equations (8.10) and (8.11), confirm that you obtained the correct result in question 8.3.

There are two special cases of equations (8.10) and (8.11) that should be considered. Firstly, if $n = 1$, we would have

$$y = x \tag{8.13}$$

and

$$dy/dx = 1x^0$$

$$= 1 \tag{8.14}$$

since $x^0 = 1$ for any value of x. This is entirely reasonable since, going back to Chapter 2, you should be able to see that equation (8.13) is a straight line through the origin with a gradient of one.

Secondly, if $n = 0$, we would have

$$y = x^0$$

$$= 1 \tag{8.15}$$

and

$$dy/dx = 0x^{-1}$$

$$= 0 \tag{8.16}$$

since any number multiplied by zero is also zero. Again this is entirely reasonable since equation (7.16) states that y equals one for all values of x. Thus a graph of y against x would be a horizontal straight line passing through $y = 1$ and this would have a gradient of zero. The important point to remember here is that the derivative of a constant is always zero.

Moving on to other standard forms in Table 8.2, the next result states that

$$\alpha = \sin(\theta) \tag{8.17}$$

has a gradient given by

$$d\alpha/d\theta = \cos(\theta). \tag{8.18}$$

Note that, since the variables here are α and θ rather than y and x, we must differentiate α with respect to θ to give a gradient in the form $d\alpha/d\theta$. The reasonableness of the result given by equations (8.17) and (8.18) should become clear if you look at Figure 8.4 in which the sine function and its derivative are shown. The upper half of this figure shows the sine function plotted as a function of angle in radians. Note that this function begins with a large positive gradient, becomes flat at $\pi/2$ radians, has a large negative slope at π radians, becomes flat again at $3\pi/2$ radians and finally becomes steeper and positive again as a complete rotation is performed at 2π radians. This therefore gives rise to a derivative of the general form shown in the lower half of Figure 8.4. This function certainly looks very like the cosine function, as predicted by equation (8.18), but is it really exactly the same? The answer is yes, provided the sine function is given as a function of the angle in radians. Thus, whenever trigonometric ratios are used in calculus, it should always be assumed that angles are measured in radians rather than degrees.

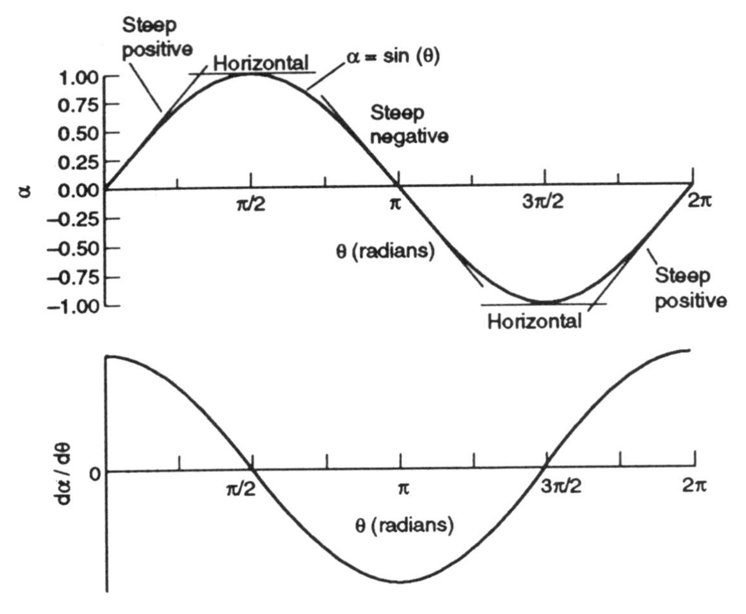

Figure 8.4 The sine function and its derivative. Note that at an angle of 0 radians the slope is steep and positive, at an angle of $\pi/2$ radians the slope is zero, at an angle of π radians the slope is steep and negative, etc. The leads to a derivative of the form shown below.

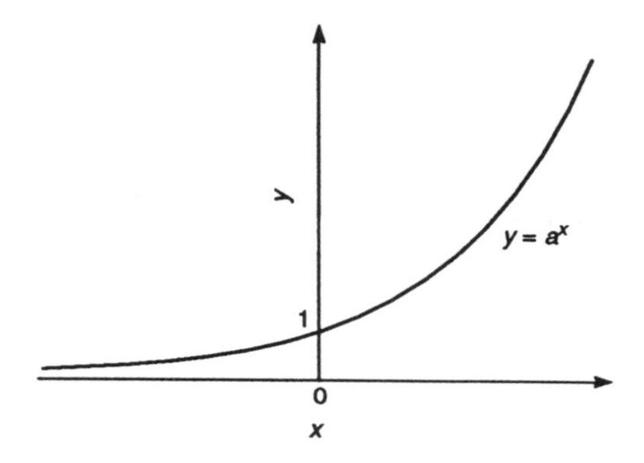

Figure 8.5 The function $y = a^x$. Note that both the function and its gradient increase continuously from small, positive, values on the left to much larger values on the right.

Skipping over the other trigonometric cases in Table 8.2, the exponential case is particularly interesting. Figure 8.5 should help to remind you what functions of the type

$$y = a^x \tag{8.19}$$

look like if a is a positive constant. The point is that these functions are always positive and increase consistently as you move to increasing values of x. They also always have a positive gradient which becomes increasingly steep with increasing x. It follows from this that the derivative of a function like equation (8.19) has a very similar appearance to the original function. In fact the number e (= 2.71...), and only this number, has the very special property that, if

$$y = \mathrm{e}^x \tag{8.20}$$

then

$$\mathrm{d}y/\mathrm{d}x = \mathrm{e}^x. \tag{8.21}$$

Question 8.5 Find the derivatives of the following functions. Be careful to get the notation correct.

(i) $y = x^{20}$. Differentiate y with respect to x.
(ii) $w = \mathrm{e}^z$. Differentiate w with respect to z.
(iii) $w = x^z$. Differentiate w with respect to x, assuming z is constant.

8.6 MORE COMPLICATED EXPRESSIONS

The trouble with standard forms is that real problems are rarely that simple. In practice, the types of expression that need to be differentiated do not appear in Table 8.2. For example, if we wished to differentiate

$$y = e^x + x^{20} - \sin(x) \tag{8.22}$$

how would we do it? The problem is that, whilst we now know how to find the derivative of each of the individual terms, how do we differentiate the whole expression? The answer is very simple indeed. All we do is differentiate the terms individually and add or subtract the results as appropriate. Thus, equation (8.22) gives

$$dy/dx = e^x + 20x^{19} - \cos(x) \tag{8.23}$$

since the derivative of e^x is e^x, the derivative of x^{20} is $20x^{19}$ and the derivative of $\sin(x)$ is $\cos(x)$. The next problem is that an expression might be multiplied by a constant. How would you differentiate

$$\alpha = 32\cos(\theta). \tag{8.24}$$

Now, the cosine function is in Table 8.2 so we can find its derivative ($= -\sin(\theta)$). The function $32\cos(\theta)$ is simply 32 times steeper than the function $\cos(\theta)$ and therefore

$$d\alpha/d\theta = -32\sin(\theta). \tag{8.25}$$

In other words, if a function is multiplied by a constant, simply multiply the derivative by the same constant.

Question 8.6 Differentiate the following:

(i) $y = x^{10} + \tan(x)$, with respect to x.
(ii) $y = x^2 + x^3 + x^4$, with respect to x.
(iii) $w = 75\,e^z$, with respect to z.
(iv) $\alpha = 10\sin(\theta) + 13\cos(\theta)$, with respect to θ.

The rules for addition and for multiplication by a constant enable us to differentiate something useful at last. What is the rate of increase in temperature with depth at a depth of 1000 km? We can start with one of the temperature equations introduced in Chapter 2. The temperature, T, in degrees Celsius at a depth, z, in kilometres is given approximately by

$$T = (-8.255 \times 10^{-5})z^2 + 1.05z + 1110. \tag{2.6}$$

Now, if we want to know the gradient at a particular depth, we must first obtain an expression for gradient as a function of depth. This is simply done by differentiating equation (2.6) with respect to z. This yields

$$\begin{aligned} dT/dz &= 2(-8.255 \times 10^{-5})z + 1.05 \\ &= (-1.651 \times 10^{-4})z + 1.05. \end{aligned} \tag{8.26}$$

To obtain the gradient at $z = 1000\,\text{km}$ we now substitute this depth into equation (8.26) which gives

$$\begin{aligned} \mathrm{d}T/\mathrm{d}z &= (-1.651 \times 10^{-4} \times 10^{3}) + 1.05 \\ &= -0.1651 + 1.05 \\ &= 0.88\,^{\circ}\mathrm{C\,km^{-1}}. \end{aligned} \qquad (8.27)$$

Thus, at a depth of 1000 km, the earth's temperature increases by 0.88 °C for each additional kilometre of depth. Note that, to solve problems of this type, you must differentiate first and then substitute the depth. You cannot solve this problem by first substituting the depth of 1000 km into equation (2.6) and then differentiating afterwards.

Question 8.7 Equation (2.8),

$$T = az^4 + bz^3 + cz^2 + dz + e \qquad (2.8)$$

with values for the constants of $a = -1.12 \times 10^{-12}$, $b = 2.85 \times 10^{-8}$, $c = -0.000\,310$, $d = 1.64$ and $e = 930$, gave a slightly different temperature versus depth function. Use this equation to evaluate the gradient at a depth of 1000 km. Compare to the answer which was obtained above using equation (2.6).

8.7 THE PRODUCT RULE AND THE QUOTIENT RULE

The next complication that must be dealt with is the case of a function which is the product of two simpler functions. For example,

$$y = x^2 \sin(x), \qquad (8.28)$$

which is the product of the function x^2 with the function $\sin(x)$. We need a general rule for dealing with these cases so, instead of the specific problem of equation (8.28) I will use the more general expression

$$y = uv \qquad (8.29)$$

where u and v are any two functions. For example, in equation (8.28)

$$u = x^2 \qquad (8.30)$$

and

$$v = \sin(x). \qquad (8.31)$$

The **product rule** states that

$$\mathrm{d}y/\mathrm{d}x = u(\mathrm{d}v/\mathrm{d}x) + v(\mathrm{d}u/\mathrm{d}x). \qquad (8.32)$$

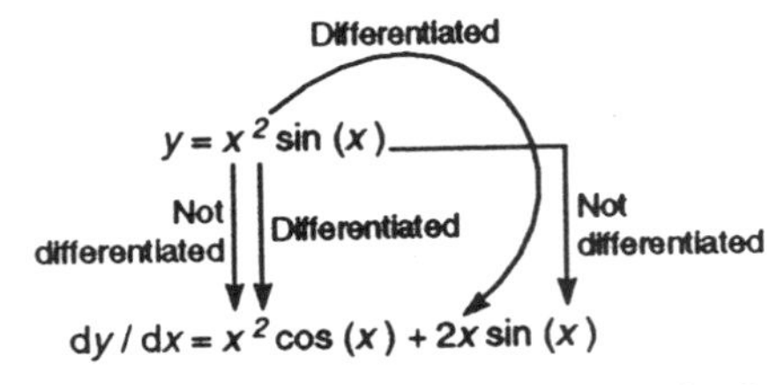

Figure 8.6 The product rule applied to equation (8.28). The first term of the answer is the first function times the derivative of the second. The second term is the second function times the derivative of the first.

Thus, there are two terms to the answer. The first term consists of the first function multiplied by the derivative of the second whilst the second term consists of the second function times the derivative of the first. To make this clearer, the example of equation (8.28) is illustrated in Figure 8.6.

More formally, from equations (8.30) and (8.31) it follows that the derivative of the first function is

$$du/dx = 2x \tag{8.33}$$

and the derivative of the second function is

$$dv/dx = \cos(x). \tag{8.34}$$

Substituting equations (8.30), (8.31), (8.33) and (8.34) into equation (8.32) yields

$$dy/dx = x^2 \cos(x) + 2x \sin(x). \tag{8.35}$$

Question 8.8 Find the derivatives of:

(i) $\alpha = x^2 e^x$.
(ii) $y = 3w^2 \sin(w)$.
(iii) $z = x\cos(x) + x^3 \tan(x)$.
(iv) $B = 3\sigma^4 \ln(\sigma) + 17\sigma^2$.

Question 8.9 The **quotient rule** states that if

$$y = u/v$$

where u and v are functions of x, the derivative of y with respect to x is given by

$$\frac{dy}{dx} = \frac{v(du/dx) - u(dv/dx)}{v^2}.$$

Use this rule to find the derivative of $y = x^4/\sin(x)$. (Hint: let $u = x^4$ and $v = \sin(x)$.)

8.8 THE CHAIN RULE

Sometimes it is necessary to make a substitution to find the derivative of an expression. If this is done then the **chain rule** should be used to obtain the required solution. Take the case of

$$y = \sin(x^2). \tag{8.36}$$

To differentiate this, make the substitution

$$z = x^2 \tag{8.37}$$

which gives

$$y = \sin(z). \tag{8.38}$$

Now, it is a simple matter to differentiate equation (8.38) using the table of standard forms. The result is

$$\mathrm{d}y/\mathrm{d}z = \cos(z). \tag{8.39}$$

However, note that this is the derivative of y with respect to z. If we could differentiate equation (7.36) the result would be the derivative of y with respect to x. So, how do we change from one to the other? The chain rule must be used. This states that

$$\mathrm{d}y/\mathrm{d}x = (\mathrm{d}y/\mathrm{d}z)(\mathrm{d}z/\mathrm{d}x) \tag{8.40}$$

for the case of the symbols we have used in this example. The left-hand side of this is the derivative we require. The right-hand side has two parts, the first of which is the derivative we have found (i.e. equation (8.39)). To complete the calculation, therefore, we need to know $\mathrm{d}z/\mathrm{d}x$. This can easily be determined by differentiating equation (8.37). Thus

$$\mathrm{d}z/\mathrm{d}x = 2x \tag{8.41}$$

which, after substituting into equation (8.40), gives

$$\begin{aligned} \mathrm{d}y/\mathrm{d}x &= \cos(z)2x \\ &= \cos(x^2)2x \\ &= 2x\cos(x^2). \end{aligned} \tag{8.42}$$

With the chain rule available to us, we can find the gradient of the porosity/depth function introduced in Chapter 2, i.e.

$$\phi = \phi_0 \mathrm{e}^{-z/\lambda} \tag{2.17}$$

where ϕ is porosity, z is depth and ϕ_0 and λ are constants. To find the gradient of porosity with depth, first make the substitution

$$x = -z/\lambda \tag{8.43}$$

This leads to

$$\phi = \phi_0 \mathrm{e}^x \tag{8.44}$$

giving

$$d\phi/dx = \phi_0\, e^x. \qquad (8.45)$$

In addition, equation (8.43) gives

$$dx/dz = -1/\lambda. \qquad (8.46)$$

Finally, the chain rule for this problem is

$$d\phi/dz = (d\phi/dx)(dx/dz). \qquad (8.47)$$

This is constructed simply by writing down, on the left-hand side, the derivative we would like to know and then writing on the right-hand side the two derivatives we have found (i.e. equations (8.45) and (8.46)). The result can be checked simply by imagining that dx in the two parts of the right-hand side is 'cancelled'. The result is $d\phi/dz$ which is the same as the left-hand side. If this does not work then the chain rule has been incorrectly written down for your problem. (N.B. Although this is a good way to check the correctness of the chain rule, it must be emphasized that this 'cancellation' is a convenient fiction. Functions such as $d\phi/dx$ do not denote $d\phi$ divided by dx, they denote the gradient of ϕ as a function of x. In other words, $d\phi/dx$ is a single function, it is not one function divided by another.)

Thus, we can now substitute equations (8.45) and (8.46) into the chain rule (equation (8.47)) to give

$$\begin{aligned} d\phi/dz &= (\phi_0\, e^x)(-1/\lambda) \\ &= (-\phi_0/\lambda)\, e^{-z/\lambda}. \end{aligned} \qquad (8.48)$$

A simpler form for this expression arises by noting that equation (8.48) is very similar to equation (2.17) leading to

$$d\phi/dz = -\phi/\lambda. \qquad (8.49)$$

For example, if the surface porosity of a shale is 60% and the decay distance λ is 500 m, then the porosity and porosity gradient at a depth of 1000 m are

$$\begin{aligned} \phi &= 60 \exp(-1000/500) \\ &= 8.12\% \end{aligned} \qquad (8.50)$$

from equation (2.17), and

$$\begin{aligned} d\phi/dz &= -8.12/500 \\ &= -0.0162\%\, m^{-1} \end{aligned} \qquad (8.51)$$

from equation (8.49).

Question 8.10 Differentiate $x = \ln(y^2)$ with respect to y.

8.9 SO, WHAT IS IT FOR?

Calculus is probably one of the most important topics in the whole of applied maths. Some of the most useful applications for calculus need **differential equations** (beyond the scope of this book) or **integral calculus** (the subject of Chapter 9). Nevertheless, there are useful things that can be done with simple differential calculus. One of these follows from the fact that the derivative of a function gives the rate of change of the function. Thus, the amount by which the function changes over a given distance can easily be found. An example used in geophysics is essential in **gravity surveying**.

Gravity surveying techniques are based upon the fact that the strength of gravity at the earth's surface is increased by dense material in the subsurface and reduced by less dense materials. Hence, in principle, measurements of the strength of gravity can indicate the subsurface distribution of rocks of differing densities. However, altitude also affects the strength of gravity. Gravity is reduced as distance from the centre of the earth increases. Now, if a gravity survey is made over a mountain range, each measurement would be taken at a different altitude. Any difference in the gravity would then be due to the combination of changes in the subsurface density distribution and changes in altitude. However, the geophysical survey is intended to reveal only the subsurface density variation since there are easier ways of determining altitude! Thus, before interpreting the gravity survey, it is nomal to remove the altitude effect. This is called the **free-air correction**. Note that there are several more corrections to gravity measurements which should also be made but the details of these are beyond the scope of this book. However, the free-air correction is probably the most important correction which must be made to gravity data before they are interpreted.

So, what is this correction? We start with the formula for the acceleration due to gravity introduced in Chapter 3:

$$g = GM/r^2$$
$$= GMr^{-2}. \qquad (3.6)$$

The next step is to differentiate equation (3.6) with respect to distance, r, from the earth's centre. This gives

$$\mathrm{d}g/\mathrm{d}r = -2GMr^{-3}. \qquad (8.52)$$

Substituting

$$r = 6370\,\mathrm{km} = 6.37 \times 10^6\,\mathrm{m}$$
$$G = 6.672 \times 10^{-11}\,\mathrm{m^3\,kg^{-1}\,s^{-2}}$$
$$M = 5.95 \times 10^{24}\,\mathrm{kg}$$

then gives

$$\mathrm{d}g/\mathrm{d}r = -3.072 \times 10^{-6}\,\mathrm{s^{-2}}. \qquad (8.53)$$

In other words, the strength of gravity reduces by $3.072 \times 10^{-6}\,\mathrm{m\,s^{-2}}$ for every metre of increased height. For example, if two measurements are separated by an altitude of 350 m, there will be a difference of

$$350 \times 3.072 \times 10^{-6} = 0.001\,08\ \mathrm{m\,s^{-2}}$$

entirely caused by the altitude difference. This correction should be added to the measurement with the higher altitude before a comparison is made.

Question 8.11 A gravity survey gave the following results:

Measurement	*Gravity* ($m\,s^{-2}$)	*Height above datum (m)*
1	9.78350	0
2	9.78012	1100
3	9.78196	500

Using the altitude correction given in equation (8.53) determine whether or not the differences in these measurements are significant assuming that the measurements are accurate to within $0.000\,02\ \mathrm{m\,s^{-2}}$.

8.10 HIGHER DERIVATIVES

The whole point about differentiation is that it produces a new function which is the gradient of the initial function. Obviously, therefore, it is quite possible to differentiate the result again to produce the gradient of the gradient function (Figure 8.7). Figure 8.7(a) shows a simple function which is differentiated to give the function shown in Figure 8.7(b) Figure 8.7(c) then shows the result of differentiating the function shown in Figure 8.7(b). Following line B through the three parts of the figure should make this clear. In Figure 8.7(a) line B passes through a part of the function which has a negative gradient. Line B therefore passes through Figure 8.7(b) at a point where this derivative has a negative value. The function in Figure 8.7(b) itself has a negative slope at the point given by line B and therefore its derivative (Figure 8.7(c)) is also negative at this point.

The only thing that is remotely difficult about forming these **higher derivatives** is the notation. This is best explained using a simple example. Take the case of

$$\chi = \sin(\psi). \tag{8.54}$$

Differentiation produces

$$\mathrm{d}\chi/\mathrm{d}\psi = \cos(\psi). \tag{8.55}$$

The derivative of this new function is then found by differentiating again. The derivative of cosine is − sine and therefore further differentiation of this result

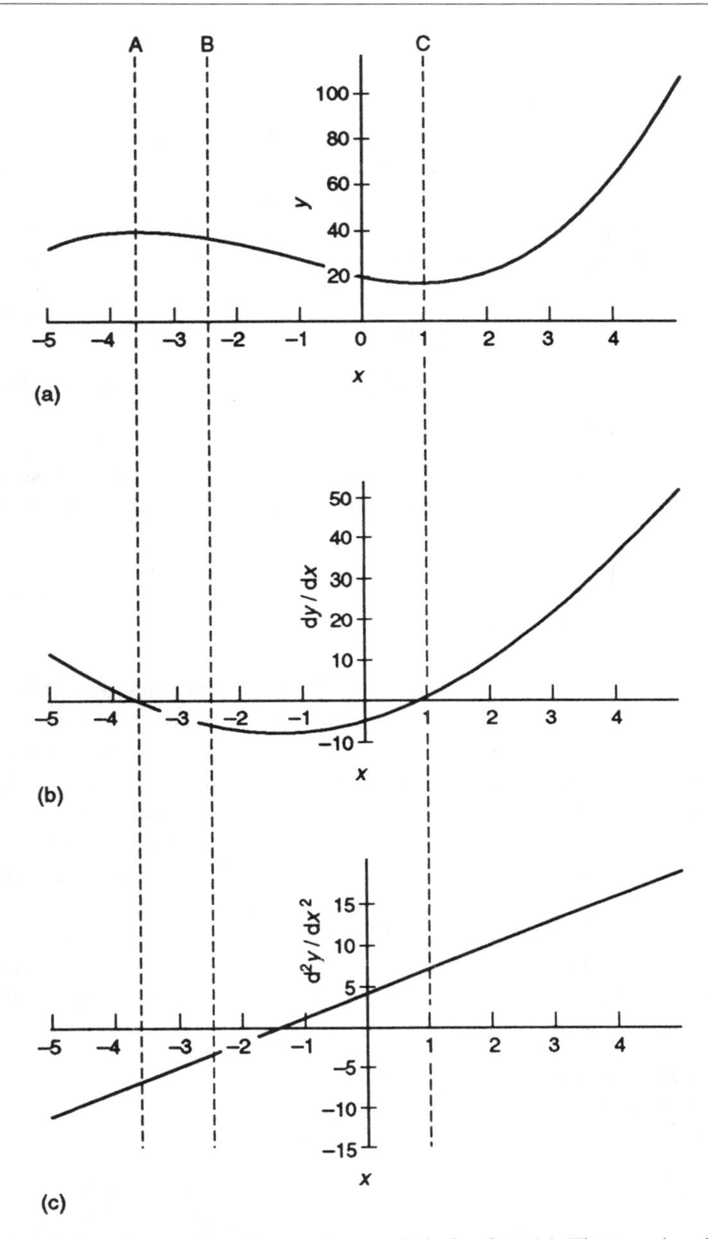

Figure 8.7 A function and its first and second derivative. (a) The starting function. (b) The derivative of the function shown in (a). (c) The derivative of the function shown in (b). This is therefore the second derivative of the function shown in (a). Lines A, B and C allow comparisons to be made at specific values of x. Note in particular that lines A and C correspond to a maximum and a minimum respectively for the function shown in (a).

gives

$$d^2\chi/d\psi^2 = -\sin(\psi) \tag{8.56}$$

which is called the second derivative of equation (8.54). Note that the fact that this result has been obtained by differentiating the function twice is indicated simply by placing superscripted 2s as indicated on the left-hand side. This process can be carried on indefinitely. For example, differentiation of − sine yields − cosine and therefore equation (8.56) after differentiation gives the third derivative

$$d^3\chi/d\psi^3 = -\cos(\psi) \tag{8.57}$$

where the superscripted 3s are introduced since the original function has now been differentiated three times.

Question 8.12 If $y = 3x^2 + 2x - 1$, obtain dy/dx and d^2y/dx^2. Also write down the third and fourth derivatives making sure to get the notation correct.

8.11 MAXIMA AND MINIMA

The higher derivatives can be used to find the **maxima** and **minima** of functions. Two such points are shown on Figure 8.7. Lines A and C pass through points on the function shown in Figure 8.7(a) which are a maximum (all adjacent points are lower) and a minimum (all adjacent points are higher) respectively. Now, at these points the gradient is zero. Thus, lines A and C indicate points in Figure 8.7(b) where the derivative is zero. Note also that, to the left of the maximum, the gradient is positive and to the right it is negative. Thus, the gradient at this point is decreasing and the gradient of Figure 8.7(b) is therefore negative in this area. This means that the second derivative (Figure 8.7(c)) must be negative at the point corresponding to the maximum. Conversely, at the point corresponding to the minimum in Figure 8.7(a), the second derivative must be positive. To summarize:

1. At maxima and minima the first derivative is zero.
2. At maxima the second derivative is negative.
3. At minima the second derivative is positive.

This ability to determine the maxima and minima of a function allows a large class of new problems to be solved using differential calculus. These are called **optimization** problems and apply whenever we wish to determine the optimum value of some parameter. An example follows that might appear a little complicated. However, each of the separate steps has been covered above and you should be able to follow these steps with occasional glances at the preceding sections.

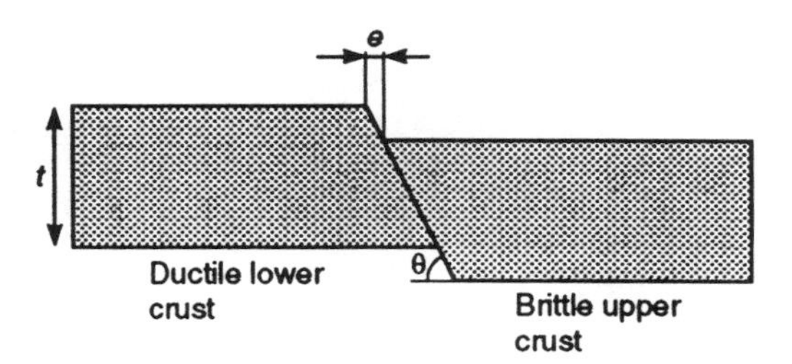

Figure 8.8 Brittle upper crust of thickness t is extended by an amount e. This deformation is accommodated by movement along a fault with a dip of θ. What fault dip would allow this to be done with the least possible amount of work?

Large faults in the subsurface typically occur at an angle of around 60° to the horizontal. One theory about why this might be so is because this angle minimizes the amount of work that has to be done on the crust in order to achieve a given amount of extension. To test this theory we need to find an expression for the amount of work done as a function of fault dip. We then find the angle for which this function is a minimum. In other words we find the optimum angle from the point of view of minimizing work done. Figure 8.8 shows the essential elements of this problem.

The amount of work necessary to achieve this extension will obviously increase if the fault length increases since there is then a larger surface for friction to act on. In addition, the amount of work done will also increase as the amount of movement on the fault increases. Now, if the fault dip is small, the length of fault needed to penetrate a given thickness of crust is large (Figure 8.9). On the other hand, if the fault becomes steep, the amount of movement along the fault needed to achieve a given extension must increase (Figure 8.10). Thus, if the fault is too shallow a lot of work must be done because the fault is long but if the fault is too steep a lot of work must be done because a lot of movement is needed. We need to find an intermediate angle where these two effects are both relatively small.

The simplest assumptions that we can make about this problem are that the work expended, w, is proportional to the length of the fault, L, and also proportional to the amount of fault movement, m. Thus

$$w = \alpha L m \tag{8.58}$$

where α is a positive constant whose precise value is unimportant in what follows. Expressions for L and m in terms of fault dip are now needed. These follow from simple trigonometry which gives

$$L = t/\sin(\theta) \tag{8.59}$$

and

$$m = e/\cos(\theta). \tag{8.60}$$

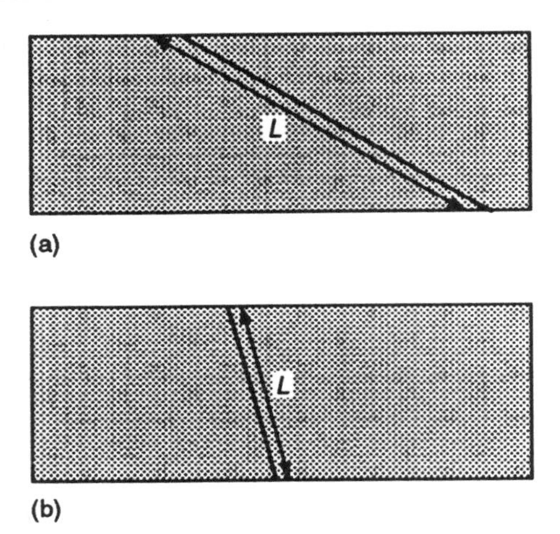

Figure 8.9 Fault length, L, to penetrate a given thickness of crust increases as dip decreases. The shallow dipping fault in (a) has a length much greater than (b) which has a much steeper fault dip.

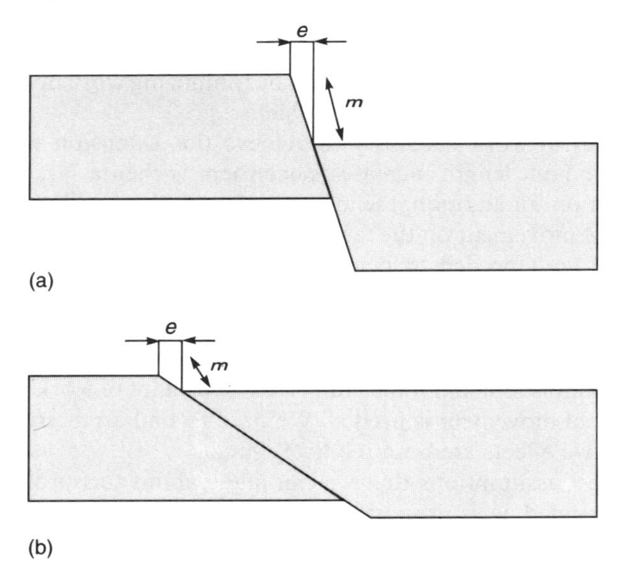

Figure 8.10 The distance moved along the fault, m, for a given amount of extension, e, depends upon fault dip. (a) The fault is steep and m is large compared to (b) where the fault is more shallow.

Thus, combining equations (8.58)–(8.60) gives

$$w = \alpha et/[\cos(\theta)\sin(\theta)]. \tag{8.61}$$

This is our expression for work done as a function of fault dip. All that remains is to find the angle θ for which w is minimized. From the earlier discussion, this occurs where the derivative of equation (8.61) is zero. The derivative can be found using the quotient rule introduced in section 8.7. For this example, use the substitutions

$$u = \alpha et \tag{8.62}$$

and

$$v = \cos(\theta)\sin(\theta) \tag{8.63}$$

from which it follows that

$$\mathrm{d}u/\mathrm{d}\theta = 0 \tag{8.64}$$

and

$$\mathrm{d}v/\mathrm{d}\theta = \cos^2(\theta) - \sin^2(\theta) \quad \text{(from product rule).} \tag{8.65}$$

The quotient rule then gives

$$\begin{aligned}\frac{\mathrm{d}w}{\mathrm{d}\theta} &= \frac{v\,\mathrm{d}u/\mathrm{d}\theta - u\,\mathrm{d}v/\mathrm{d}\theta}{v^2} \\ &= \alpha et\frac{\sin^2(\theta) - \cos^2(\theta)}{\cos^2(\theta)\sin^2(\theta)}.\end{aligned} \tag{8.66}$$

All that now remains is to find the angles for which this is zero. Now αet is generally non-zero. Thus, if

$$\mathrm{d}w/\mathrm{d}\theta = 0.0 \tag{8.67}$$

then

$$[\sin^2(\theta) - \cos^2(\theta)] = 0.0$$

i.e.

$$\sin^2(\theta) = \cos^2(\theta). \tag{8.68}$$

One solution to equation (8.68) is

$$\theta = 45^\circ. \tag{8.69}$$

We must now determine whether this is a maximum or a minimum. Remember that both maxima and minima have a derivative of zero. However, we specifically need a minimum since we are looking for a fault dip which minimizes the work needed for extension. To test whether a fault dip of 45° produces a minimum we must first obtain the second derivative of w. This is done by differentiating equation (8.66) (this is given as an exercise later) which gives

$$\frac{\mathrm{d}^2w}{\mathrm{d}\theta^2} = \alpha et\left(\frac{4}{\cos\theta\sin\theta} + \frac{2(\sin^2\theta - \cos^2\theta)}{(\cos\theta\sin\theta)^3}\right). \tag{8.70}$$

Now, if the fault dip is 45°, this gives a value of

$$\frac{\mathrm{d}^2w}{\mathrm{d}\theta^2} = 2\alpha et \tag{8.71}$$

which is positive since α, e and t are all positive. A fault dip of 45° therefore corresponds to a minimum in w. Thus, the theory predicts that faults should occur at 45° which contradicts the observation that faults are generally considerably steeper than this.

Thus, we have a simple quantitative test of a theory for fault formation. Since observations and theory do not agree here we are forced to conclude that something is missing in our theory. The important point is that the preceding mathematics has forced us to conclude that the simple fault model is inadequate. It would be very difficult, in the absence of such an analysis, to decide whether or not the simple fault dip theory was correct. The next step would be to consider whether the model can be modified to make it more accurate or whether it should be completely abandoned. However, more complete theories of fault formation are beyond the scope of this book.

FURTHER QUESTIONS

8.13 The thickness, S, of post-rift sediment in a simple extensional basin with a good sediment supply is well approximated by

$$S = S_{\max}(1 - e^{-t/\tau})$$

where $S_{\max}$ and τ are constants and t is time since rifting ceased.

(i) Give an expression for the sediment thickness immediately after rifting (i.e. at $t = 0$).
(ii) Give an expression for the sediment thickness when t is very large.
(iii) Give an expression for the rate of sedimentation immediately after rifting.
(iv) What is the rate of sedimentation when t is very large?
(v) If $S_{\max} = 3$ km and $\tau = 50$ My, what is the numerical value of the sedimentation rate immediately after rifting?

8.14 Figure 8.11 illustrates the barred basin model of evaporite formation. Sea water of normal salinity floods through the inlet of the basin and evaporation within the basin increases this salinity to the point where halite

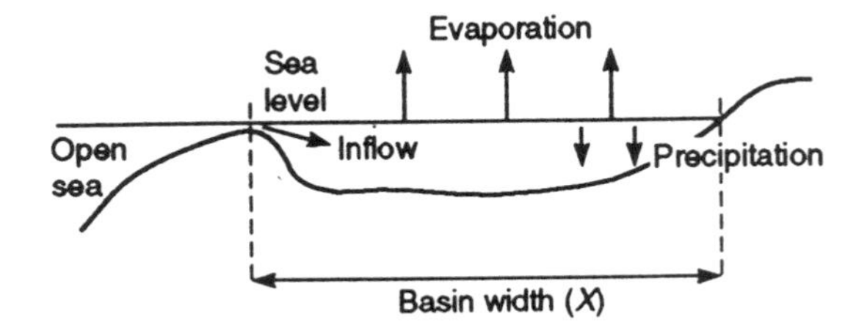

Figure 8.11 Barred basin model for evaporite formation. Sea water enters through the inlet and evaporation within the basin increases the salinity to the point where precipitation of halite and other evaporites occurs.

and other evaporites precipitate from the water and are deposited on the basin floor.

A simple mathematical theory for this process predicts that the salinity, s, within the basin increases with distance, x, from the inlet according to

$$s = s_0 \alpha X/(\alpha X - x)$$

where s_0 is the salinity of the sea water at the inlet, α is a constant and X is the basin width

(i) If water at the shore line (i.e. at $x = X$) has three times the salinity of the sea water at the inlet what is the value of α?
(ii) With this vaue for α and a sea water salinity of 30 ppm, determine the salinity at a point half-way across the basin (i.e. at $x = X/2$).
(iii) Prove that the rate of increase in salinity with distance is given by

$$\mathrm{d}s/\mathrm{d}x = s/(\alpha X - x).$$

(iv) Evaluate this salinity gradient at a point half-way across the basin if its width is 10 000 m. Hence estimate the salinity at points 1 km on either side of the basin centre.

8.15 In section 8.11 it was stated that if

$$\frac{\mathrm{d}w}{\mathrm{d}\theta} = \alpha e t \frac{\sin^2\theta - \cos^2\theta}{\cos^2\theta \sin^2\theta} \tag{8.66}$$

then

$$\frac{\mathrm{d}^2 w}{\mathrm{d}\theta^2} = \alpha e t \left(\frac{4}{\cos\theta \sin\theta} + \frac{2(\sin^2\theta - \cos^2\theta)}{(\cos\theta \sin\theta)^3} \right). \tag{8.70}$$

Prove this result using the following steps:

(i) Let $u = \alpha e t(\sin^2\theta - \cos^2\theta)$ and obtain the derivative $\mathrm{d}u/\mathrm{d}\theta$. (Hint: use the chain rule.)
(ii) Let $v = \cos^2\theta \sin^2\theta$ and obtain the derivative $\mathrm{d}v/\mathrm{d}\theta$. (Hint: write this in the form $v = (\cos\theta \sin\theta)^2$ and use both the chain rule and product rule.)
(iii) Use the quotient rule to obtain $\mathrm{d}^2 w/\mathrm{d}\theta^2$.

8.16 The thickness of a bottomset bed at the foot of a delta can often be well approximated by the expression

$$t = t_0 \exp(-x/X)$$

where t is thickness, x is distance from the bottomset bed start and t_0 and X are constants.

(i) Determine an expression for the rate of change of bed thickness.
(ii) Evaluate this expression at a point 3 km from the bottomset bed start if $X = 5$ km and $t_0 = 10$ m.

9 Integration

9.1 INTRODUCTION

Integration is simply the inverse process to differentiation. The integral of $2x$ is x^2 because the derivative of x^2 is $2x$ (this statement is not quite complete, I will discuss why later). As such, integration may seem a little abstract and not particularly useful. Nothing could be further from the truth! As I will show later, integration allows such things as the variation in density with depth in the earth to be estimated, the heat generated in the crust to be determined and the approximate volume of Mt Fuji to be calculated. However, to start with I will discuss integration as the inverse of differentiation so that I can introduce the methods and notation used for calculating integrals.

9.2 INDEFINITE INTEGRATION

Indefinite integration is the reverse process to differentiation. From this definition it follows that the integral of any given function can be found by discovering a function with the appropriate derivative. For example, if the integral of $\cos(\theta)$ is required, we need only remember that $\cos(\theta)$ can be obtained by differentiating $\sin(\theta)$. Thus, integrating $\cos(\theta)$ should give $\sin(\theta)$. Note, however, that the derivative of $\sin(\theta) + 3$ is also $\cos(\theta)$ and, in general, the derivative of $\sin(\theta) + k$ equals $\cos(\theta)$ for any constant k. Thus, the general solution to the integral of $\cos(\theta)$ is $\sin(\theta) + k$ where k is an unknown constant. This is written

$$\int \cos(\theta)\,\mathrm{d}\theta = \sin(\theta) + k. \tag{9.1}$$

The notation may look a little strange. It is, perhaps, easiest if you think of $\int$ as being a 'left-hand integration bracket' and $\mathrm{d}\theta$ as being a 'right-hand integration bracket'. Equation (9.1) then states that integrating the function between the brackets gives $\sin(\theta) + k$. The reason for this strange notation should become apparent later.

Similarly, one of the problems in Chapter 8 involved differentiating

$$y = \mathrm{e}^x + x^{20} - \sin(x) \tag{8.22}$$

to give

$$\mathrm{d}y/\mathrm{d}x = \mathrm{e}^x + 20x^{19} - \cos(x). \tag{8.23}$$

From this it follows that

$$\int (\mathrm{e}^x + 20x^{19} - \cos(x))\,\mathrm{d}x = \mathrm{e}^x + x^{20} - \sin(x) + k. \tag{9.2}$$

Equation (9.2) is entirely equivalent to equations (8.22) and (8.23). This is just a different way of writing almost the same thing! Note that the integration is performed with respect to θ in equation (9.1) and with respect to x in equation (9.2). In other words, the variable being integrated is θ in equation (9.1) and x in equation (9.2). Hence, in equation (9.1) the 'right-hand integration bracket' is $\mathrm{d}\theta$ whilst in equation (9.2) it is $\mathrm{d}x$.

Question 9.1 From Chapter 8 it should be clear that the derivative of $5\,\mathrm{e}^{2\alpha}$ with respect to α is $10\,\mathrm{e}^{2\alpha}$. Using this fact, write an integral equation similar to equations (9.1) or (9.2). Do not forget the constant and remember that you are now integrating with respect to α and thus the function to be integrated is enclosed between $\int$ and $\mathrm{d}\alpha$.

From the preceding discussion it should be clear that a table of standard integrals can be produced from a table of standard derivatives simply by swapping around the columns. Thus, reproducing Table 8.2 and rearranging the columns gives a table of integrals (Table 9.1) in which k is an integration constant. Note that the first and third examples in Table 9.1 are not particularly simple as they stand and therefore of limited use as standard integrals. The first expression in Table 9.1 can be improved upon by starting

Table 9.1 A table of integrals formed by swapping the columns in Table 8.2 and adding an integration constant

y	$\int y\,\mathrm{d}x$
nx^{n-1}	$x^n + k$
$\cos(x)$	$\sin(x) + k$
$-\sin(x)$	$\cos(x) + k$
$1/\cos^2(x)$	$\tan(x) + k$
e^x	$\mathrm{e}^x + k$
$1/x$	$\ln(x) + k$

with the fact that if

$$y = \frac{x^{n+1}}{n+1} \tag{9.3}$$

then

$$\frac{dy}{dx} = x^n \tag{9.4}$$

from which it follows that

$$\int x^n \, dx = \frac{x^{n+1}}{n+1} + k. \tag{9.5}$$

Note, however, that if $n = -1$ this expression does not give a sensible answer (try it!). In this particular case use the result from Table 9.1 that $\int x^{-1} \, dx = \ln(x) + k$.

Question 9.2 Prove that equation (9.4) follows from equation (9.2) using the rules of differentiation given in Chapter 8.

Similarly, it is easy to show that

$$\int \sin(x) \, dx = -\cos(x) + k. \tag{9.6}$$

Using these two results in place of the first and third in Table 9.1 gives a table of standard integrals (Table 9.2). This list is a very short one. There are large books containing standard integrals for many other functions.

Question 9.3 Using Table 9.2 evaluate:

(i) $\int \cos(x) \, dx$.
(ii) $\int \xi^{10} \, d\xi$. (Hint: this is solved using the first result in Table 9.2.)

Table 9.2 A table of standard integrals formed by simplifying some of the results in Table 9.1

y	$\int y \, dx$
x^n	$x^{n+1}/(n+1)$
$\cos(x)$	$\sin(x) + k$
$\sin(x)$	$-\cos(x) + k$
$1/\cos^2(x)$	$\tan(x) + k$
e^x	$e^x + k$
$1/x$	$\ln(x) + k$

9.3 INTEGRATION OF MORE COMPLEX EXPRESSIONS

As with differentiation, the problem with standard forms is that real problems are rarely that simple. We need a few rules for dealing with more complex cases. Some of these are similar to those for differentiation. For example, if a function is multiplied by a constant, the integral of the function is multiplied by the same constant. Thus, in general,

$$\int a f(x)\,\mathrm{d}x = a \int f(x)\,\mathrm{d}x \tag{9.7}$$

where a is a constant and $f(x)$ is any function of x. For instance

$$\begin{aligned}\int 3\sin(x)\,\mathrm{d}x &= 3\int \sin(x)\,\mathrm{d}x \\ &= 3(-\cos(x)+k) \quad \text{(from Table 9.2)} \\ &= -3\cos(x)+c \end{aligned} \tag{9.8}$$

where c is an integration constant (which happens to equal $3k$ but this is unimportant since k itself is not known). Another rule is that the integral of the sum of several functions is simply given by the sum of the integrals. Thus

$$\int [f(x)+g(x)-h(x)]\,\mathrm{d}x = \int f(x)\,\mathrm{d}x + \int g(x)\,\mathrm{d}x - \int h(x)\,\mathrm{d}x \tag{9.9}$$

where $f(x)$, $g(x)$ and $h(x)$ are any functions of x. Note that, although three functions are shown in equation (9.9), this principle applies to any number of functions. An example of this is

$$\begin{aligned}\int [\cos(x)-x^2]\,\mathrm{d}x &= \int \cos(x)\,\mathrm{d}x - \int x^2\,\mathrm{d}x \\ &= \sin(x) - (x^3/3) + k. \end{aligned} \tag{9.10}$$

Note that here I have made use of the fact that the sum of two constants is also a constant.

These two rules can be combined to solve problems such as

$$\begin{aligned}\int [\mathrm{e}^\theta + 5\theta^{19}]\,\mathrm{d}\theta &= \int \mathrm{e}^\theta\,\mathrm{d}\theta + 5\int \theta^{19}\,\mathrm{d}\theta \\ &= \mathrm{e}^\theta + 5(\theta^{20}/20) + k \\ &= \mathrm{e}^\theta + (\theta^{20}/4) + k. \end{aligned} \tag{9.11}$$

Question 9.4 Using the above rules and Table 9.2, integrate $(10/x)+\cos(x)$ with respect to x.

These are all the integration rules that we need in this book although you will find textbooks full of many more rules and tips. Integration is, in practice, frequently very difficult or even impossible. However, with the additional rules you can find in other texts and the published lists of standard forms, many problems can be solved. I would prefer to spend time on applications of integration but, before we can look at these, there is one more aspect of integration which must be understood.

9.4 DEFINITE INTEGRATION

The integration discussed above was called indefinite integration since, because of the existence of the unknown constant, the final answer is not completely determined. **Definite integration**, on the other hand, produces a single, often numerical, answer. The simplest way to explain definite integration is with an example such as

$$y=\int_1^2 x\,\mathrm{d}x. \tag{9.12}$$

This is very similar to earlier examples apart from the numbers above and below the integration sign. These numbers are called the **integration limits.** Ignore these for now and proceed as before to give

$$y=\left[\frac{x^2}{2}+k\right]_1^2 \tag{9.13}$$

where the limits have been written outside the square brackets just to remind us that they are still there. So what do we do with them? The answer is to substitute each in turn in place of x and to subtract the results. Thus

$$\begin{aligned} y&=\left(\frac{2^2}{2}+k\right)-\left(\frac{1^2}{2}+k\right)\\ &=2-0.5+k-k\\ &=1.5 \end{aligned} \tag{9.14}$$

which is the final answer. Notice that the constant, k, cancels out and therefore, in definite integration only, it can be ignored. Also note that the lower limit result is subtracted from the upper limit result. One more example before you try one yourself:

$$\begin{aligned} \int_0^{\pi/2}\cos(\alpha)\,\mathrm{d}\alpha&=[\sin(\alpha)]_0^{\pi/2} \quad \text{(from Table 9.2)}\\ &=\sin(\pi/2)-\sin(0) \quad \text{(after substitution of limits)}\\ &=1.0-0.0\\ &=1.0. \end{aligned} \tag{9.15}$$

Remember that calculus is always performed using radians rather than degrees and that $\pi/2$ radians is the same as 90°.

Question 9.5 Evaluate the following definite integral:

$$y = \int_0^1 4\,\mathrm{e}^x\,\mathrm{d}x.$$

9.5 INTEGRATION AS A SUMMATION

Definite integration can be used to solve many problems. For example, how would you go about measuring the volume of Mt Fuji (Figure 9.1)? One way to do this is to split the mountain into a series of thin slices (Figure 9.2), each of which has a volume which can be calculated. Now, the volume of a slice is simply its area times its thickness, Δz. I have used Mt Fuji for simplicity since, being a very symmetric volcanic structure, the slices are in fact well approximated by small circular discs. Thus, since the area of the disc in Figure 9.2 is πr^2 (where r is the disc radius), the volume, ΔV, is given by

$$\Delta V = \pi r^2 \Delta z. \tag{9.16}$$

Figure 9.1 A print of Mount Fuji by Hokusai from the series *36 views of Mount Fuji* published during the 1830s. Apart from a little exaggeration of the steepness of the upper slopes, this picture captures the fundamental shape of this volcano very accurately.

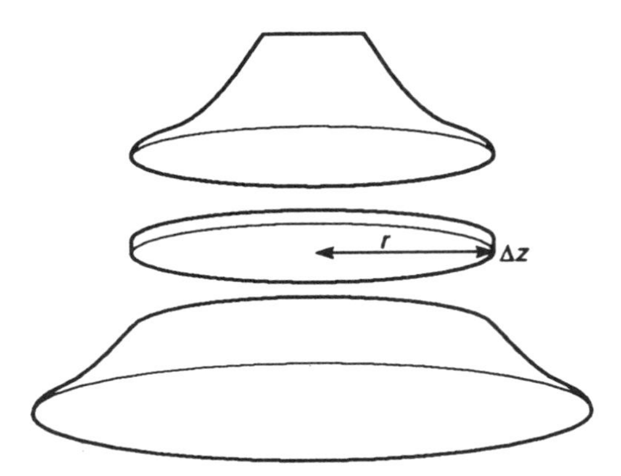

Figure 9.2 Estimating the volume of Mt Fuji by summing the volume of a large number of small slices.

The volume of the entire mountain is then approximately equal to the sum of the volumes of all these slices, i.e.

$$V = \sum_{i=1}^{N} \Delta V_i \tag{9.17}$$

$$= \sum_{i=1}^{N} \pi r_i^2 \Delta z \tag{9.18}$$

where ΔV_i and r_i are the volume and radius respectively of the ith disc and there are N discs from the mountain base to its peak.

So, how many discs do we need? A small number of thick discs would produce a mountain whose sides were made of large steps. As the discs become thinner, and their number greater, the sides would become gradually smoother. Thus, as the discs become thinner more discs are needed but the result becomes more accurate. To get a good result we should use a very large number of very thin discs. To get the best possible result, we should use an infinite number of infinitely thin discs. If this is done then Δz is replaced by an infinitely thin $\mathrm{d}z$ (should sound familiar, cf. Chapter 8) and, to indicate that N has become infinite, the summation sign is replaced by an $\int$ sign. Equation (9.18) then becomes

$$V = \int_{\text{bottom}}^{\text{top}} \pi r^2 \, \mathrm{d}z. \tag{9.19}$$

The 'bottom' and 'top' indicate that the first disc is at the mountain base and the last is at the mountain top. The point I am driving at is that definite integration can be interpreted as finding the sum of an infinite number of infinitely small quantities. Incidentally, the integration sign is derived from

Figure 9.3 The shape of Mt Fuji modelled by equation (9.20).

the old-fashioned 'long s' (e.g. the King James Bible of 1611 was printed with sentences such as, 'For this cauſe left I thee in Crete, that thou ſhouldeſt ſet in order the things that are wanting', Paul 1.5.) and is used because integration is a special form of summation.

To solve equation (9.19) we require a definite form for how the radius varies with height up the mountain as well as a definite height for 'bottom' and 'top'. From a map of Mt Fuji it can be shown that, to a good approximation,

$$r^2 = \frac{400z}{3} - \frac{800\sqrt{z}}{\sqrt{3}} + 400\,\text{km}^2 \tag{9.20}$$

where z goes from 0 at the base to 3 km at the top. Thus, at the mountain base, $z = 0$ and equation (9.20) gives a radius of 20 km. At the mountain top, $z = 3$ and equation (9.20) gives a radius of zero. Figure 9.3 shows the form of this expression for predicting the mountain shape. Compare it to Figure 9.1 which, despite a small amount of artistic licence exaggerating the upper slopes, shows the general shape of Mt Fuji very well. Substituting equation (9.20) into equation (9.19) then gives

$$\begin{aligned}
V &= \int_0^3 \pi\left[\frac{400z}{3} - \frac{800\sqrt{z}}{\sqrt{3}} + 400\right]\text{d}z \\
&= \pi\int_0^3 \frac{400z}{3}\,\text{d}z - \pi\int_0^3 \frac{800\sqrt{z}}{\sqrt{3}}\,\text{d}z + \pi\int_0^3 400\,\text{d}z \\
&= \pi\left[\frac{400z^2}{6} - \frac{800z^{1.5}}{1.5\sqrt{3}} + 400z\right]_0^3 \\
&= \pi(600 - 1600 + 1200) \\
&= 200\pi = 628\,\text{km}^3
\end{aligned} \tag{9.21}$$

which is the approximate volume of Mt Fuji. From this, we could estimate the age of Mt Fuji if we knew the rate of production of volcanic material. Alternatively, the average rate of production could be estimated if some other measure of volcano age could be found.

Another example is worth going through before you attempt something similar. In structural geology there are several different ways of measuring strain (i.e. the amount of deformation). One such quantity is called the strain, s, and is given by

$$s = \Delta L / L_{\text{i}} \tag{9.22}$$

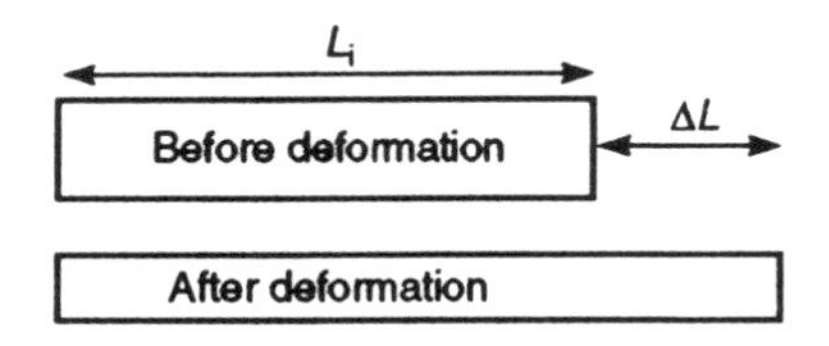

Figure 9.4 A material is deformed such that its initial length, L_i, is increased by an amount ΔL.

where L_i is the initial length of a strained material and ΔL is the amount by which the deformation increases the length (Figure 9.4).

An alternative is to consider the final strain as resulting from a large number of small deformation episodes. In the limit, the deformation can be thought of as resulting from an infinite number of infinitely small extensions. In each episode the length is increased by an amount dL and there are an infinite number of these such that the length increases from an initial length L_i to a final length L_f. Thus the resulting strain measurement, called the total natural strain, ε, is given by

$$\begin{aligned} \varepsilon &= \int_{L_i}^{L_f} \frac{\mathrm{d}L}{L} \\ &= [\ln(L)]_{L_i}^{L_f} \\ &= \ln(L_f/L_i). \end{aligned} \tag{9.23}$$

This can be compared to a third method of quantifying strain called the stretch, S, which is simply the final length divided by the initial length. From this it follows that

$$\varepsilon = \ln(S) \tag{9.24}$$

i.e. the total natural strain is simply the natural logarithm of the stretch.

Question 9.6 Heat is generated by radioactivity in the crust at a rate measured in microwatts per cubic metre ($\mu\mathrm{W\,m^{-3}}$) or, occasionally, kilowatts per cubic kilometre ($\mathrm{kW\,km^{-3}}$) (note that these two units are actually the same, i.e. $1\,\mu\mathrm{W\,m^{-3}} = 1\,\mathrm{kW\,km^{-3}}$). For example, each cubic kilometre of continental upper crust generates about 1 kW of heat (which is about the same as a typical electric room heater). This heat generation rate actually decreases with depth until, below the crust, hardly any heat is generated in this way.

If, in a particular piece of crust 30 km thick, the heat generation rate, Q, can be approximated by

$$Q = y/20\,\mathrm{kW\,km^{-3}} \tag{9.25}$$

where y is the distance from the base of the crust:

(i) Determine the heat generation rate at 0, 10, 20 and 30 km from the crust base. Sketch a graph of the result.
(ii) Write down an expression for the heat generated in a small box-shaped volume Δz kilometres thick whose other dimensions are 1 km by 1 km. (Hint: what is the volume of such a box?)
(iii) Give an expression for the approximate heat generated by a column of such boxes by writing an equation similar to (9.18).
(iv) Write down the expression that results from allowing the thickness of these crust volumes to tend to zero.
(v) Evaluate the resulting equation which gives the rate of flow of heat from each square kilometre at the surface.
(vi) What area would be needed to generate as much heat as a small sized power station (say 100 MW)?

9.6 INTEGRATING DISCONTINUOUS FUNCTIONS

The difficulty with real problems is that they are often not well approximated by the simple functions considered up to here. In particular, especially in geological problems, there are often sharp discontinuities where the properties of interest change dramatically. A good example is the density profile of the earth (Figure 9.5) which has sharp changes at various depths such as the mantle/core boundary.

This profile has been determined using a large number of different geophysical measurements. In particular, one important constraint on possible density profiles is that they must be consistent with the known mass of the earth (see Chapter 3 on how this is known). If the average density of the

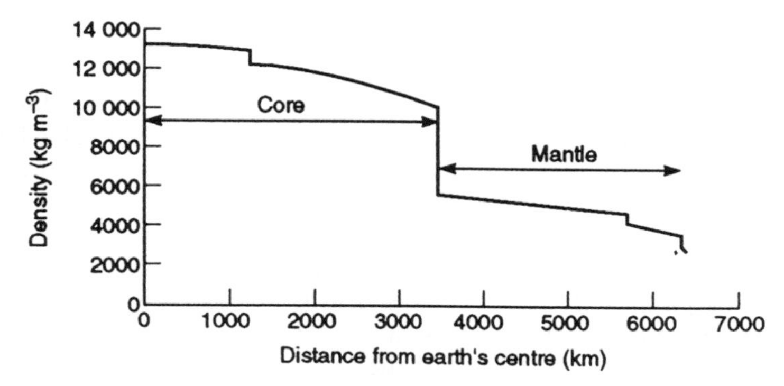

Figure 9.5 The density profile of the earth as a function of distance from the centre. Note the very sharp changes in density particularly at the core/mantle boundary.

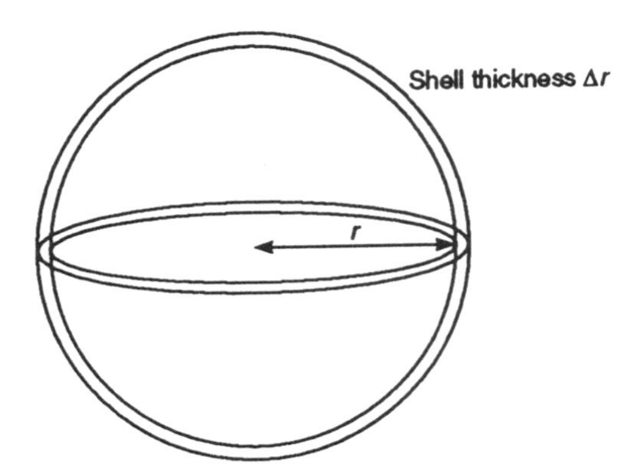

Figure 9.6 A spherical shell of radius r and thickness Δr. The surface area of a sphere is $4\pi r^2$ and thus the volume of this shell is $4\pi r^2 \Delta r$.

proposed profile is too high then the mass will also be too high and vice versa. The total mass is found by first considering the mass of a thin spherical shell of thickness Δr and radius r (Figure 9.6). The volume of a thin shell is approximately its surface area times its thickness. Thus, since the surface area of a sphere is $4\pi r^2$, the volume is $4\pi r^2 \Delta r$ and the mass is $4\pi r^2 \rho \Delta r$ where ρ is the shell density. This leads to a mass estimate of

$$M = \sum_{i=1}^{N} 4\pi r_i^2 \rho_i \Delta r \tag{9.26}$$

where there are N shells in total and ρ_i is the density of the ith shell. The accuracy of this expression increases as the number of shells increases and the shell thickness decreases until, in the limit of infinitely many infinitely thin shells, it becomes the integral

$$M = \int_0^R 4\pi r^2 \rho \, dr \tag{9.27}$$

where R is the radius of the earth.

All we need now is an expression for how density varies with radius. For simplicity we can approximate the density by a single constant value for the core and a single constant value for the mantle. Let us assume densities of 11 000 and 4500 kg m^{-3} respectively. How do we handle the change-over at the core/mantle boundary radius of 3480 km? Simple, we integrate using a density of 11 000 kg m^{-3} from the centre to the boundary to give the mass of the core. A separate calculation using a density of 4500 kg m^{-3} can then be used to integrate from the boundary to the surface to give the mantle mass. The total mass is then just the sum of the two. Thus

$$M = \int_0^{3.48 \times 10^6} 4\pi r^2 \times 11\,000 \, dr + \int_{3.48 \times 10^6}^{6.371 \times 10^6} 4\pi r^2 \times 4500 \, dr$$

$$= 44\,000\pi[r^3/3]_0^{3480} + 18\,000\pi[r^3/3]_{3480}^{6371}$$

$$= (46\,077 \times (3.48 \times 10^6)^3) + 18\,850((6.371 \times 10^6)^3 - (3.48 \times 10^6)^3)$$

$$= (1.94 \times 10^{24}) + (4.08 \times 10^{24})$$

$$= 6.02 \times 10^{24}\,\text{kg} \tag{9.28}$$

which compares quite well to the mass calculated in Chapter 3 of 5.95×10^{24} kg. The discrepancy would disappear if a density function closer to that shown in Figure 9.5 were used instead of the simple two-density model used above.

FURTHER QUESTIONS

9.7 The earth has the shape of a slightly flattened sphere, i.e. the polar radius is slightly less than the equatorial radius. Figure 9.7 is an exaggeration of this to illustrate the point. The precise shape is one in which sections cut parallel to the equator are circular whilst sections through the poles are elliptical. Such a shape is called an ellipsoid.

In an ellipsoid, the radius, r, of a circle of constant latitude at a distance z from the equator is given by

$$r^2 = r_e^2(1 - (z^2/r_p^2)). \tag{9.29}$$

(i) What is the volume of the disc illustrated in the figure in terms of its radius, r, and thickness, Δz?
(ii) Write an expression for the approximate volume of the earth by assuming it is filled with a stack of N such discs.
(iii) Rewrite the expression for the case of an infinite number of infinitely thin discs. Evaluate the resulting integral. (Hint: you will need to integrate from $z = -r_p$ to $z = r_p$. The result will be an algebraic expression rather than a simple number.)

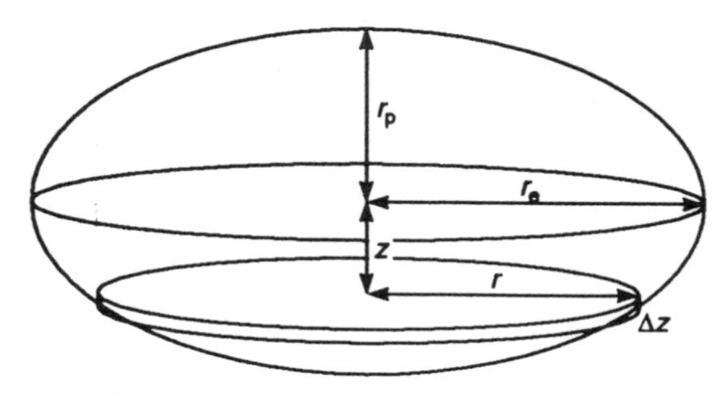

Figure 9.7 The earth is an ellipsoidal shape whose equatorial radius, r_e, is slightly greater than its polar radius r_p. A thin disc is shown which is parallel to the equator, has a radius, r, and thickness Δz and is a vertical distance, z, from the equator.

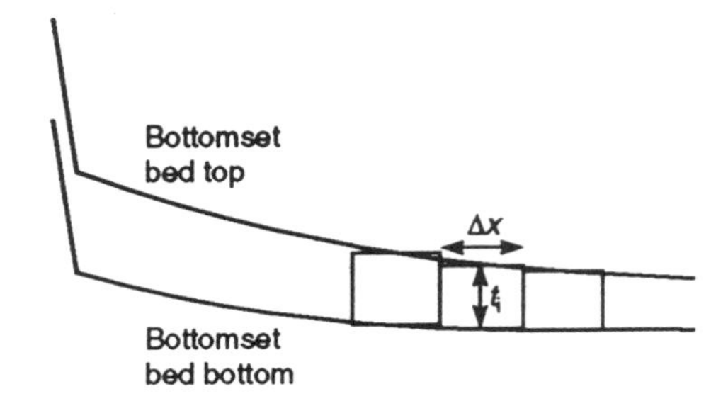

Figure 9.8 Approximating the bottomset bed by a series of rectangular elements of thickness t_i and width Δx.

(iv) If the equatorial radius of the earth is 6378 km and the polar radius is 6357 km, what is the volume?

(v) How does this compare to the volume for a sphere of radius equal to the average of the polar and equatorial values?

9.8 The thickness of a bottomset bed at the foot of a delta can often be well approximated by the expression

$$t = t_0 \exp(-x/X)$$

where t is thickness, x is distance from the bottomset bed start and t_0 and X are constants.

(i) Imagine approximating this sedimentary bed in cross-section by a series of rectangles of height t_i and width Δx (Figure 9.8). What is the area of each rectangle?

(ii) Now write down an approximate expression for the cross-sectional area of the entire bottomset bed.

(iii) By considering the limiting case of an infinite number of infinitesimally thick rectangles, write down and evaluate an integral equation giving the total cross-sectional area. (Hint: $\int e^{ax}\, dx = (1/a)e^{ax} + k$)

(iv) If the present-day rate of sediment supply is $10\,\mathrm{m^2\,y^{-1}}$, $X = 5\,\mathrm{km}$ and $t_0 = 1\,\mathrm{m}$, estimate the time taken to form the bed assuming the sediment supply rate has not altered through time.

Appendix – Answers

This section gives outline answers to all questions in this book. Note, however, that your answers should be much more complete and self-contained. The very first question which follows is more complete and should give you the general idea of how to set out your answers.

CHAPTER 1

1.1 The age of sediments in a lake bed is approximated by

$$\text{Age} = k \times \text{Depth} \tag{1.1}$$

where $k = 1500\,\text{y m}^{-1}$. Thus, if depth = 1 m, then

$$\text{Age} = 1500 \times 1 = 1500\,\text{years.}$$

If depth = 2 m, then

$$\text{Age} = 1500 \times 2 = 3000\,\text{years.}$$

If depth = 5.3 m, then

$$\text{Age} = 1500 \times 5.3 = 7950\,\text{years.}$$

Similarly, if $k = 3000\,\text{y m}^{-1}$ and depth = 1 m, then

$$\text{Age} = 3000 \times 1 = 3000\,\text{years.}$$

If depth = 2 m, then

$$\text{Age} = 3000 \times 2 = 6000\,\text{years.}$$

If depth = 5.3 m, then

$$\text{Age} = 3000 \times 5.3 = 15\,900\,\text{years.}$$

1.2 (i) $5^6 = 15\,625$, (ii) $5^8 = 390\,625$, (iii) x^5, (iv) Depth^5, (v) $T_0^{12} = 10^{12}$.
1.3 (i) 1×10^3, (ii) 2×10^3, (iii) 2.5×10^3, (iv) 2.523×10^3, (v) 2.3×10^7,

(vi) 7×10^9.
1.4 (i) 1×10^{-3}, (ii) 2×10^{-3}, (iii) 2.5×10^{-3}, (iv) 2.523×10^{-3}, (v) 2.3×10^{-6}, (vi) 7×10^{-9}.
1.5 1 year = 365.26 days × 24 hours × 3600 seconds = 31.6 million seconds. Therefore 31.6 gigaseconds = 31.6 billion seconds = 1000 × 31.6 million seconds = 1000 years. In scientific notation this is 3.16×10^{10} seconds.
1.6 100 ppm.
1.7 (i) 4×10^9, (ii) 2.65×10^9, (iii) 1×10^7, (iv) 2.35×10^{14}.
1.8 $5.51 \times 10^3 = 5510\,\text{kg}\,\text{m}^{-3}$.
1.9 300 years.
1.10 8.7×10^{-7}.
1.11 Rate of increase in mass ΔM is

$$\begin{aligned} 6 \times 10^5\,\text{kg}\,\text{day}^{-1} &= 6 \times 10^5 \times 365.26\,\text{kg}\,\text{y}^{-1} \\ &= 2.19 \times 10^8\,\text{kg}\,\text{y}^{-1}. \end{aligned}$$

Increase in mass is

$$\begin{aligned} \Delta M \times A_e &= (2.19 \times 10^8) \times (4.5 \times 10^9) \\ &= 2.19 \times 4.5 \times 10^{17} \\ &= 9.86 \times 10^{17}\,\text{kg}. \end{aligned}$$

Increase as a fraction of the earth's total mass is

$$\frac{9.86 \times 10^{17}}{5.95 \times 10^{24}} = 1.66 \times 10^{-7} = 166\,\text{ppb}.$$

1.12 $1.08 \times 10^{21}\,\text{m}^3$.
1.13 (i) $A = w/v_s$. (ii) 1.25×10^8 years = 125 million years.

CHAPTER 2

2.1 1.025 My.
2.2 $500\,\text{y}\,\text{m}^{-1}$.
2.3 Your rate should be around $3800\,\text{y}\,\text{m}^{-1}$ and the lake dried out about 545 ky ago.
2.4 $\tau = \tau_0 + mP$. The graph should be a straight line of gradient m and intercept τ_0.
2.5 2, − 10 and 6 respectively.
2.6 Equation (2.6) predicts $T = 3403\,°\text{C}$. Equation (2.8) predicts $T = 3648\,°\text{C}$. These are both less than the true temperature of 3700 °C but equation (2.8) is much closer than equation (2.6).
2.7 You answers should be close to 0.45.
2.8 0.15.
2.9 0.42.
2.10 25.

2.11 (i) Around 23 y cm^{-1}. (ii) Around 1.9 y cm^{-1}. (iii) About 1490 years.
2.12 4.4 ppm.
2.13 The specimen age is around 23 My.
2.14 The ore body lies between 0.5 and 2.0 km since, in this area, the gravity strength is slightly higher than expected.
2.15 (i) See Figure A.1. (ii) p_0 is the accumulation rate at zero depth, Z is the

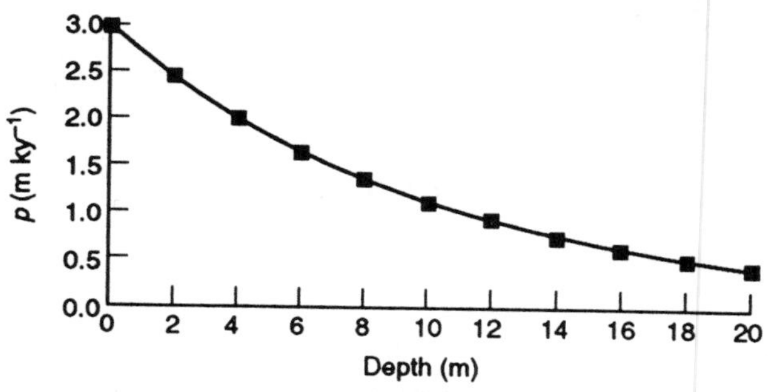

Figure A.1 Question 2.15.

depth at which accumulation has fallen off to approximately $\frac{1}{3}$ of its maximum value (more precisely, the depth at which accumulation is p_0/e).

CHAPTER 3

3.1 $k = \text{Age/Depth} = 3000/3 = 1000\ \text{y m}^{-1}$.
3.2 Age of top = Age − k × Depth = 60 000 − (5000 × 10) = 10 000 years.
3.3 $\dfrac{w}{x} = \dfrac{3y/(4z)}{2y/(4z)} = \dfrac{3y}{2y} = \dfrac{3}{2} = 1.5.$
3.4 (i) $5x + 10y$, (ii) $5x + 11y$, (iii) $5.5x + 11y$, (iv) $5ax + 10ay$, (v) $x^2 - 4y^2$, (vi) $x^2 + 4xy + 4y^2$.

3.5

$$\begin{aligned}\text{Depth} &= (1/k)(\text{Age} - \text{Age of top})\\ &= (1/k) \times \text{Age} - (1/k)(\text{Age of top})\\ &= \frac{\text{Age}}{k} - \frac{\text{Age of top}}{k}.\end{aligned}$$

3.6 $3a(2x + y)$.
3.7 See Figure A.2.
3.8 Both roots equal 1.
3.9 Depth = 2170 km or depth = 10 549 km.
3.10 The equations are

$$\rho = \rho_g[1 - (V_p/V)]$$

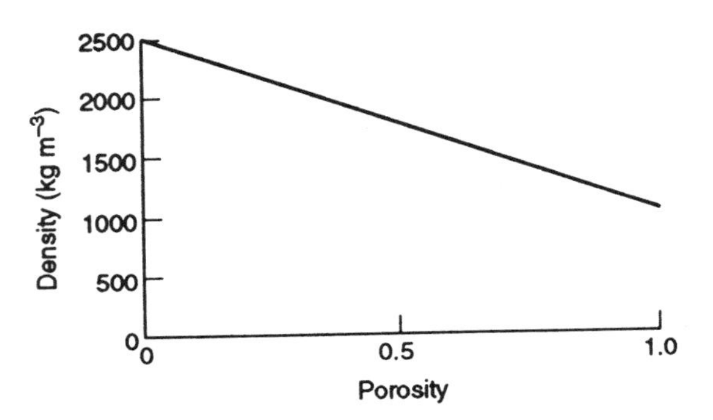

Figure A.2 Question 3.7.

and

$$\rho = M/V$$

where ρ is density, ρ_g is the density of the grains making up the rock, V_p is the volume occupied by pore space, V is the total volume and M is mass. Since these equations both equal ρ we can write immediately

$$\rho_g[1-(V_p/V)] = M/V.$$

Dividing by $[1-(V_p/V)]$ gives

$$\rho_g = M/\{V[1-(V_p/V)]\}$$

which can be multiplied through to yield

$$\rho_g = M/[V - V_p]$$

which is the required answer.

If total volume is $0.11\,m^3$ and porosity is 0.32, the pore space has a volume of $0.32 \times 0.11 = 0.0352\,m^3$. Substituting $V = 0.11\,m^3$, $V_p = 0.0352\,m^3$ and $M = 205\,kg$ then gives

$$\begin{aligned}\rho_g &= 205/[0.11 - 0.0352] \\ &= 2741\,kg\,m^{-3}\end{aligned}$$

which is the grain density, whilst the average density is simply

$$\begin{aligned}\rho &= M/V \\ &= 205/0.11 \\ &= 1864\,kg\,m^{-3}\end{aligned}$$

3.11 10 days/100 = 0.1 days = 2 hours 24 minutes.

3.12 (i) $b = -ax - (c/x)$.

(ii) From (i) $b^2 = a^2x^2 + (c^2/x^2) + 2ac$, hence $b^2 - 4ac = a^2x^2 + (c^2/x^2) - 2ac$.

(iii) Since $(ax - (c/x))^2 = a^2x^2 + (c^2/x^2) - 2ac$.
(iv) From (ii) and (iii) $\sqrt{(b^2 - 4ac)} = (ax - c/x)$.
Hence, using (i), $-b + \sqrt{(b^2 - 4ac)} = ax + (c/x) + ax - (c/x) = 2ax$ and therefore

$$\frac{-b + \sqrt{(b^2 - 4ac)}}{2a} = 2ax/2a = x.$$

CHAPTER 4

4.1 About 15 million years.
4.2 11.04 ppb.
4.3 $a = 1.1$, $b = 2$, $c = 3$.
4.4 $a \approx -1 \times 10^{-4}$, $b \approx 1$, $c \approx -2000$. Thus

$$z \approx \frac{-1 + \sqrt{(1 - 0.8)}}{-2 \times 10^{-4}} \quad \text{(use a calculator to get } \sqrt{0.2}\text{)}$$

$$= \frac{-1 + 0.45}{-2 \times 10^{-4}}$$

$$= 5500/2 = 2250\,\text{km}.$$

4.5 $G = gr^2/M$.
4.6 At zero depth $T = T_0$ which is sensible.
4.7 (i) $\ln(t_0) = \ln(t) + (x/X)$.
(ii) $\ln(t_0) = \ln(5) + (1/X)$ and $\ln(t_0) = \ln(0.1) + (4/X)$. Hence

$$\ln(5) + (1/X) = \ln(0.1) + (4/X)$$

gives

$$X = 3/\ln(50) = 0.767\,\text{km}.$$

(iii) $t_0 = t/\exp(-x/X)$
$= 5/\exp(-1/0.767)$ (when evaluated at $x = 1$ km)
$= 18.4$ m.

4.8 If $T = ar^4 + br^2 + c$ and $T = 4300\,°\text{C}$ at $r = 1260$ km then

$$4300 = a \times 1260^4 + b \times 1260^2 + c$$
$$= (2.52 \times 10^{12})a + (1.59 \times 10^6)b + c. \qquad (1)$$

Similarly, if T 1900 °C at $r = 5660$ km then

$$1900 = (1.03 \times 10^{15})a + (3.20 \times 10^7)b + c \qquad (2)$$

and, if $T = 1150\,°\text{C}$ at $r = 6260$ km then

$$1150 = (1.54 \times 10^{15})a + (3.92 \times 10^7)b + c. \qquad (3)$$

These three equations can be solved, simultaneously, to yield

$$a = -6.82 \times 10^{-13} \qquad b = -5.59 \times 10^{-5} \qquad c = 4391.$$

At the earth's centre, $r = 0$, the predicted temperature is just c, i.e. 4391 °C, which is a sensible value. At the earth's surface, $r = 6360$ km, the predicted temperature is 1015 °C which is unrealistically high.

4.9 a has units of °C km^{-2}, b has units of °C km^{-1} and c has units of °C.

4.10 (i) Wrong since the units of (depth × rate) are not ky.
(ii) Wrong since the units of (rate/depth) are not ky.
(iii) Wrong since the units of 'Depth of top' are not ky.
(iv) Wrong since at a depth of zero this gives 'Age = Age of bottom bed' rather than 'Age = Age of top bed.'

CHAPTER 5

5.1 (a) $\beta = 60°$, $b = 10.1$ cm, $c = 11.5$ cm.
(b) $c = 2.4$ cm, $\alpha = 25°$, $\beta = 135°$.

5.2 Church is about 2.2 km from the transmitter. Outcrop is 3.1 km from the church and 2.6 km from the transmitter.

5.3 (i) 180° is half a full rotation. Therefore 180° is equivalent to $\frac{1}{2} \times 2\pi = \pi$ radians, i.e. about 3.14 radians.
(ii) Similarly, 90° is $\pi/2$ radians.
(iii) 270° is $3\pi/2$ radians.
(iv) Finally, 100° is

$$\frac{100}{360} = \frac{5}{18} \text{ of a complete rotation which is } \frac{5}{18} \times 2\pi \text{ radians} = 1.75 \text{ radians.}$$

5.4 Assume the adjacent side has length x and the hypotenuse has a length of $2x$. Thus, the length of the opposite side is $\sqrt{(4x^2 - x^2)} = \sqrt{3}x$. The definitions for sine, cosine and tangent then give

$$\sin(\text{angle}) = \text{Opposite/Hypotenuse} = \frac{\sqrt{3}x}{2x} = \frac{\sqrt{3}}{2}.$$

$$\cos(\text{angle}) = \text{Adjacent/Hypotenuse} = \frac{x}{2x} = \frac{1}{2}$$

$$\tan(\text{angle}) = \text{Opposite/Adjacent} = \frac{\sqrt{3}x}{x} = \sqrt{3}.$$

If you chose the opposite side to have length x instead of the adjacent side, your results will be $1/2$, $\sqrt{3}/2$ and $1/\sqrt{3}$ for sin, cos and tan respectively.

5.5 141.9 m.

5.6 $\tan(\text{dip}) = 45/110 = 0.409$. Thus, $\text{dip} = \tan^{-1}(0.409) = 22.2°$.

5.7 (a) 40°, (b) 74.6°, (c) 2.28 km.

5.8 $a^2 = b^2 + c^2 - 2bc\cos(90) = b^2 + c^2 - (2bc \times 0) = b^2 + c^2$.
5.9 Angles are 108.2°, 49.5° and 22.3°.
5.10 (i) $\beta = 119.7°$, $\gamma = 20.3°$, $a = 3.7$ km.
(ii) $\beta = 74.6°$, $\gamma = 65.4°$, $c = 2.8$ km.
(iii) $\gamma = 80°$, $b = 4.0$ km, $c = 4.6$ km.
5.11 $\tan(\theta_c) = \text{Opposite/Adjacent} = -4/-4 = 1$.
$\tan(\theta_d) = \text{Opposite/Adjacent} = -8/8 = -1$.
5.12 From equation (5.30), $\cos(\alpha) = \tan(\theta')/\tan(\theta)$. Therefore

$$\alpha = \cos^{-1}[\tan(\theta')/\tan(\theta)] = 48.2°.$$

5.13 (i) scalar, (ii) vector, (iii) scalar, (iv) scalar, (v) vector.
5.14 See Figure A.3 for vectors **a**, **b**, **c** and **d**. Vector **e** is of zero length.

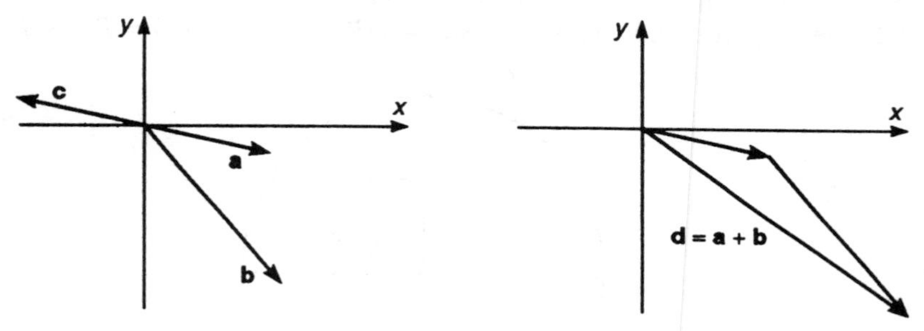

Figure A.3 Question 5.14.

5.15 From Figure A.4,

$$\cos\theta = x/a, \quad \text{therefore} \quad x = a\cos\theta.$$
$$\sin\theta = y/a, \quad \text{therefore} \quad y = a\sin\theta.$$

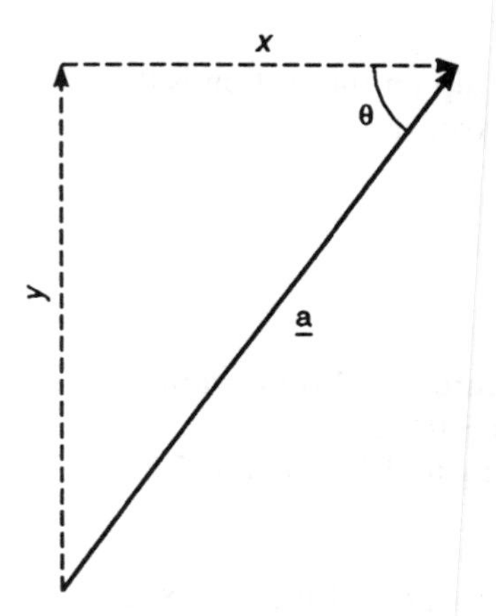

Figure A.4 Question 5.15.

5.16 Fault (i) has components $x = 10\cos(60) = 5$ and $y = 10\sin(60) = 8.66$.
Fault (ii) has components $x = 5\cos(65) = 2.11$ and $y = 5\sin(65) = 4.53$.
Fault (iii) has components $x = 12\cos(45) = 8.49$ and $y = 12\sin(45) = 8.49$.
Thus, the components of the total slip vector are

$$x = 5 + 2.11 + 8.49 = 15.6$$

and

$$y = 8.66 + 4.53 + 8.49 = 21.68.$$

In terms of length and direction, this is a slip vector of length

$$\sqrt{(15.6^2 + 21.68^2)} = 26.71 \text{ m}$$

with a dip of $\tan^{-1}(21.68/15.6) = 54.26°$.
5.17 (i) 0.966, (ii) 0.932, (iii) 26.6°, (iv) 0.794, (v) 114.6.
5.18 See Figure A.5. From this, $\cos(\phi) = r/R$ and therefore $r = R\cos(\phi)$.

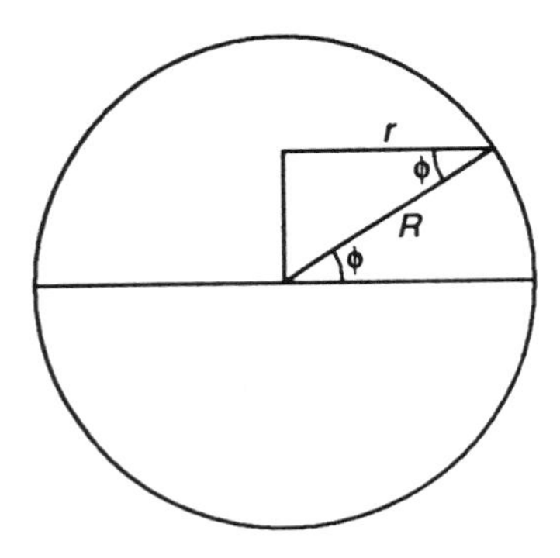

Figure A.5 Question 5.18.

5.19 436 m.
5.20 The apparent dip on the section is 8.53°. Section is 32° from a true dip section. Hence, the true dip is 10°.
5.21 352° E of N.

CHAPTER 6

6.1 Graphs on Figure A.6. From graph (iii), gradient = – 0.082 and intercept = 7.5. Hence $b = 0.082$.
6.2 Mass is 55 000 kg and foot area is 2 m^2.
6.3 See Figure A.7.
6.4 See Figure A.8.
6.5 Dip of 10° is 10 × (90 – 10)/90 = 8.89 cm from plot centre. Similarly, dip of 20° is 10 × (90 – 20)/90 = 7.78 cm from plot centre. Thus two points with

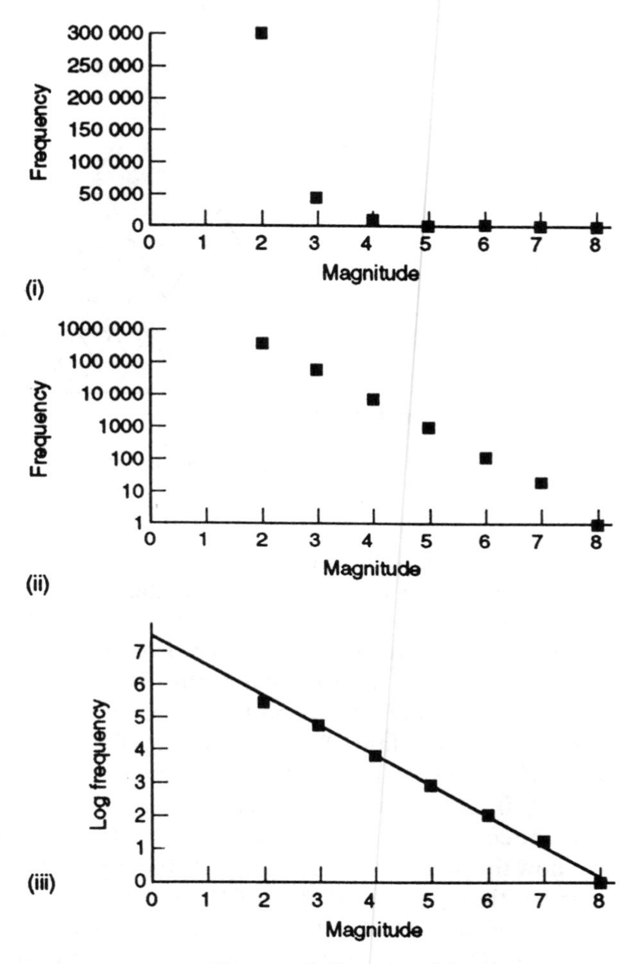

Figure A.6 Question 6.1.

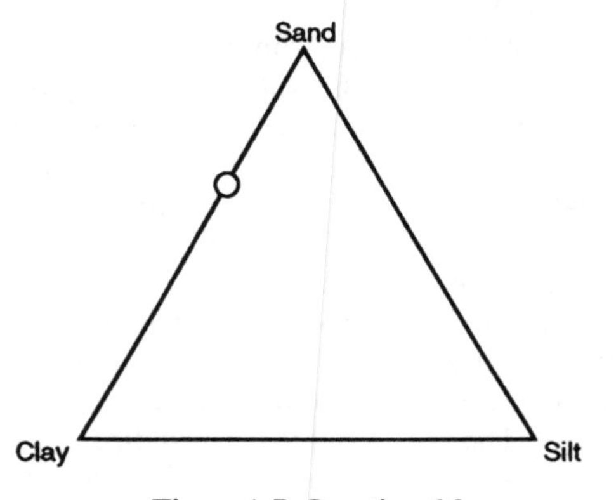

Figure A.7 Question 6.3.

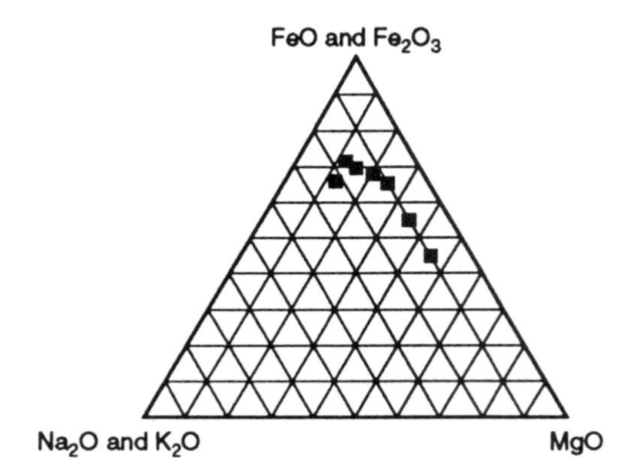

Figure A.8 Question 6.4.

same azimuth and dips of 10° and 20° will be 1.11 cm apart. Repeating this for the case of dips of 70° and 80° gives the same answer, i.e. these two points are also 1.11 cm apart.

6.6 Dip of 10° is $10\tan(40) = 8.39$ cm from plot centre. Similarly, dip of 20° is $10\tan(35) = 7.00$ cm from plot centre. Thus two points with same azimuth and dips of 10° and 20° will be 1.39 cm apart.

Dip of 70° is $10\tan(10) = 1.76$ cm from plot centre. Similarly, dip of 80° is $10\tan(5) = 0.87$ cm from plot centre. Thus two points with same azimuth and dips of 70° and 80° will be 0.89 cm apart.

In the equal interval case these two pairs of points had identical separations, but now the points with same azimuth and dips of 70° and 80° are much closer together than points with dips of 10° and 20°.

6.7 Dip of 10° is $14.14\sin(40) = 9.09$ cm from plot centre. Similarly, dip of 20° is $14.14\sin(35) = 8.11$ cm from plot centre. Thus two points with same azimuth and dips of 10° and 20° will be 0.98 cm apart.

Dip of 70° is $14.14\sin(10) = 2.46$ cm from plot centre. Similarly, dip of 80° is $14.14\sin(5) = 1.23$ cm from plot centre. Thus two points with same azimuth and dips of 70° and 80° will be 1.26 cm apart.

The points with same azimuth and dips of 70° and 80° are further apart than points with dips of 10° and 20°. This contrasts with the equal interval case where the separations are identical and the equal angle case in which the 70° and 80° points are closer together than the 10° and 20° points.

6.8 A triangular plot.

6.9 A straightforward x–y plot of area as a function of age.

6.10 A polar plot.

6.11 (i) The poles should lie close to a great circle.

(ii) The pole dip = 90° − bed dip.
The pole azimuth = 180° + bed dip direction.

Thus, bed dip towards 40° E of N has pole azimuth 210° E of N. Bed dip of 20° has a pole dip of 70°.

(iii) See Figure A.9.

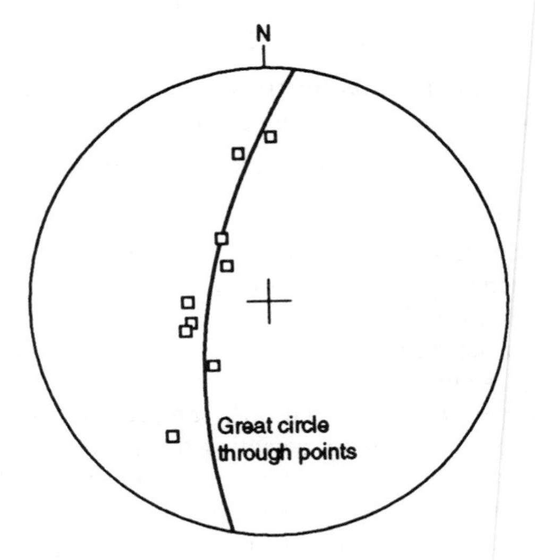

Figure A.9 Question 6.11.

(iv) Yes, the poles lie on, or near, a great circle.

CHAPTER 7

7.1 Possible combinations are HHH, HHT, HTH, THH, HTT, THT, TTH, TTT. Thus

No. of heads	No. of combinations
0	1
1	3
2	3
3	1

7.2 A sample.

7.3 (i) 681.8, (ii) 669, (iii) 82 665, (iv) 91 850 and 303.

7.4 Probability, $P_{0.4}$, of lying within 0.4 standard deviations is 0.311.
Probability, $P_{0.5}$, of lying within 0.5 standard deviations is 0.383.
Thus, probability, $P_{0.44}$, of lying within 0.44 standard deviations is given

by

$$P_{0.44} \approx 0.6P_{0.4} + 0.4P_{0.5}$$
$$= (0.6 \times 0.311) + (0.4 \times 0.383)$$
$$= 0.187 + 0.153 = 0.340.$$

Probability, $P_{0.9}$, of lying within 0.9 standard deviations is 0.632. Thus probability of lying between 801 and 1000 g is

$$0.5 \times (0.632 - 0.340) = 0.146.$$

7.5 Gradient = 1973 y m^{-1}, intercept = 61.8 years.

7.6 (i) The null hypothesis is that there is no significant difference between the specimen length and those of typical *Orthoceras* specimens, i.e. the specimen is an *Orthoceras.*

(ii) Yes, accept the specimen as *Orthoceras* since this length can occur by chance more often than 5% of the time

7.7 Standard error = standard deviation/$\sqrt{}$sample size = $436/\sqrt{10}$ = 138 g.
Pebbles 1 to 10, mean = 682 g, deviation from 610 g = 72 g.
Pebbles 11 to 20, mean = 582 g, deviation from 610 g = 28 g.
Pebbles 31 to 40, mean = 192 g, deviation from 610 g = 419 g.
Pebbles 51 to 60, mean = 830 g, deviation from 610 g = 220 g.
The biggest deviation is three times bigger than the standard error and the smallest is about 1/5 of the standard error.

7.8 Yes.

7.9 (i)

Range	*Frequency*	*Probability*
0–44° E of N	9	0.45
45–89° E of N	2	0.1
90–134° E of N	2	0.1
135–179° E of N	0	0.0
180–224° E of N	3	0.15
225–269° E of N	1	0.05
270–314° E of N	2	0.1
315–359° E of N	1	0.05

(ii) See Figure A.10.

(iii) There is a strong NNE trend.

7.10

$$s^2 = \frac{1}{N}\left(\sum_{i=1}^{N}(w_i - \bar{w})^2\right)$$

$$= \frac{1}{N}\left(\sum_{i=1}^{N}(w_i^2 + \bar{w}^2 - 2w_i\bar{w})\right)$$

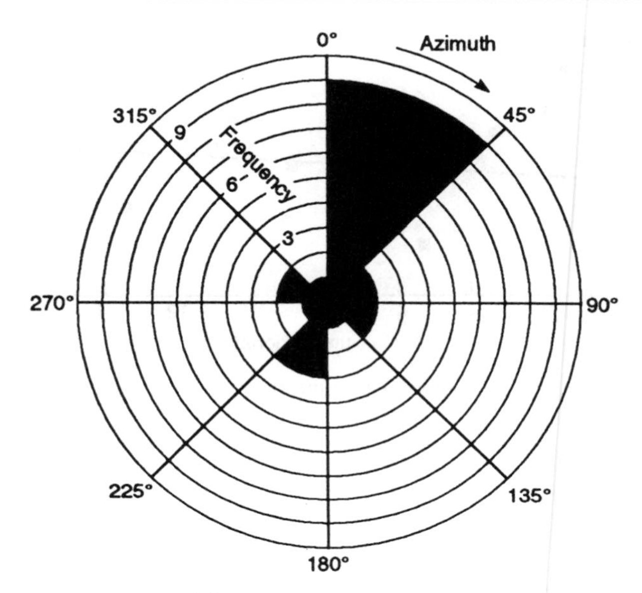

Figure A.10 Question 7.9.

$$= \frac{1}{N}\sum_{i=1}^{N} w_i^2 + \frac{1}{N}\sum_{i=1}^{N} \bar{w}^2 - \frac{1}{N}\sum_{i=1}^{N} 2w_i\bar{w}$$

$$= \frac{1}{N}\sum_{i=1}^{N} w_i^2 + \bar{w}^2 - \frac{2\bar{w}}{N}\sum_{i=1}^{N} w_i$$

$$= \overline{w^2} + \bar{w}^2 - 2\bar{w}^2$$

$$= \overline{w^2} - \bar{w}^2.$$

7.11 (i) 0.365.
(ii) This indicates a skew which is also apparent on Figure 7.1.
(iii) The population is not skewed and the skew in the sample is therefore a statistical fluctuation.

7.12 (i) % organic carbon = (0.011 × % calcium carbonate) + 0.606.
(ii) See Figure A.11.
(iii) The data are not a good fit to a straight line.

CHAPTER 8

8.1 See Figure A.12. Gradient of tangent = − 12%/0.4 km = − 30% km^{-1}.

8.2 Slope = 4 at $x = 2$. Slope = 2000 at $x = 1000$.

8.3 If

$$y = x^3 \tag{1}$$

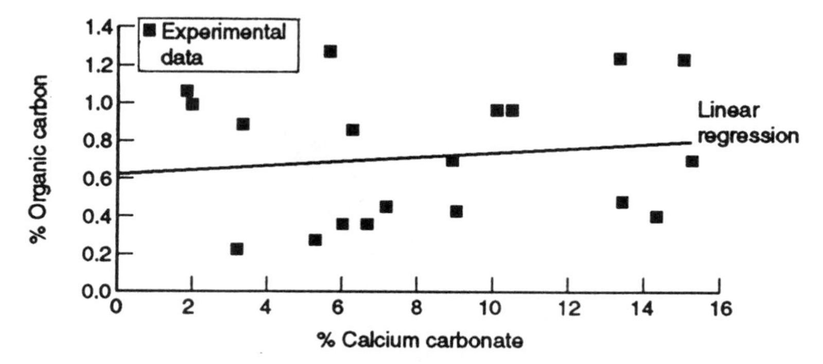

Figure A.11 Question 7.12.

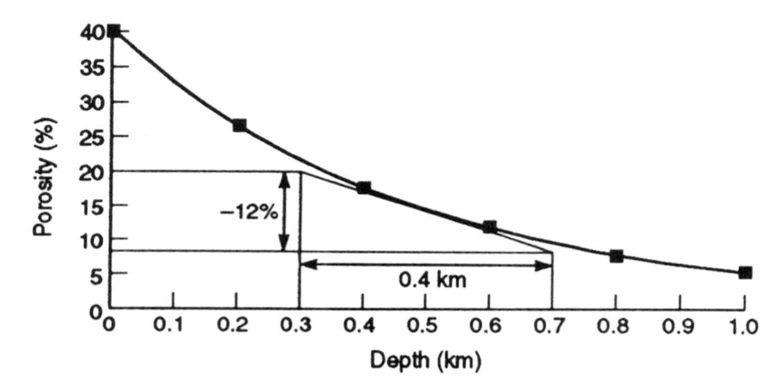

Figure A.12 Question 8.1.

then

$$y + \delta y = (x + \delta x)^3 \tag{2}$$

where δx is the small change in x caused by δy, a small change in y.

Multiplying out equation (2)

$$y + \delta y = x^3 + \delta x^3 + 3x\delta x^2 + 3x^2\delta x. \tag{3}$$

Subtracting equation (1) from equation (3)

$$\delta y = \delta x^3 + 3x\delta x^2 + 3x^2\delta x$$

which leads to

$$\frac{\delta y}{\delta x} = \delta x^2 + 3x\delta x + 3x^2 \tag{4}$$

As δx tends to zero, equation (4) becomes

$$\frac{\mathrm{d}y}{\mathrm{d}x} = 3x^2. \tag{5}$$

8.4 If

$$y = x^n \tag{8.10}$$

then

$$\frac{dy}{dx} = nx^{n-1}. \tag{8.11}$$

Thus, if

$$y = x^3$$

then

$$\frac{dy}{dx} = 3x^{3-1} = 3x^2.$$

8.5 (i) $\frac{dy}{dx} = 20x^{19}$, (ii) $\frac{dw}{dz} = e^z$, (iii) $\frac{dw}{dx} = zx^{z-1}$.

8.6 (i) $\frac{dy}{dx} = 10x^9 + \frac{1}{\cos^2(x)}$, (ii) $\frac{dy}{dx} = 2x + 3x^2 + 4x^3$, (iii) $\frac{dw}{dz} = 75\,e^z$,

(iv) $\frac{d\alpha}{d\theta} = 10\cos(\theta) - 13\sin(\theta)$.

8.7 If

$$T = az^4 + bz^3 + cz^2 + dz + e \tag{2.8}$$

then

$$\frac{dT}{dz} = 4az^3 + 3bz^2 + 2cz + d.$$

Thus, at $z = 1000$ km and for constants equal to $a = -1.12 \times 10^{-12}$, $b = 2.85 \times 10^{-8}$, $c = -0.000\,310$, $d = 1.64$ and $e = 930$,

$$\frac{dT}{dz} = 4(-1.12 \times 10^{-12}) \times 10^9 + 3(2.85 \times 10^{-8}) \times 10^6 + 2(-0.000\,310) \times 1000 + 1.64$$

$$= 1.101\,°\mathrm{C\,km^{-1}}.$$

This is significantly higher than the result from the quadratic equation.

8.8 (i) $\frac{d\alpha}{dx} = (2x + x^2)e^x$, (ii) $\frac{dy}{dw} = 3w^2\cos(w) + 6w\sin(w)$,

(iii) $\frac{dz}{dx} = \cos(x) - x\sin(x) + 3x^2\tan(x) + \frac{x^3}{\cos^2(x)}$,

(iv) $\frac{dB}{d\sigma} = 12\sigma^3\ln(\sigma) + 3\sigma^3 + 34\sigma$.

8.9

$$y = x^4/\sin(x) \tag{1}$$

$$= u/v \tag{2}$$

where

$$u = x^4 \tag{3}$$

and

$$v = \sin(x). \tag{4}$$

From equation (3)

$$\frac{du}{dx} = 4x^3. \tag{5}$$

From equation (4)

$$\frac{dv}{dx} = \cos(x). \tag{6}$$

Substituting equations (3)–(6) into

$$\frac{dy}{dx} = \frac{v(du/dx) - u(dv/dx)}{v^2}$$

gives

$$\frac{dy}{dx} = \frac{\sin(x)4x^3 - x^4\cos(x)}{\sin^2(x)}$$

$$= \frac{4x^3\sin(x) - x^4\cos(x)}{\sin^2(x)}.$$

8.10 $\frac{dx}{dy} = \frac{2}{y}$.

8.11 After altitude correction, these measurements differ by less than the measurement error. Thus, the differences are not significant.

8.12

$$\frac{dy}{dx} = 6x + 2$$

$$\frac{d^2y}{dx^2} = 6$$

$$\frac{d^3y}{dx^3} = 0$$

$$\frac{d^4y}{dx^4} = 0.$$

8.13 (i) $S = 0$, (ii) $S = S_{max}$, (iii) $\frac{dS}{dt} = \frac{S_{max}}{\tau}$, (iv) $\frac{dS}{dt} = 0$, (v) $20\ \text{m ky}^{-1}$.

8.14

$$s = s_0\alpha X/(\alpha X - x).$$

(i) At $x = X$, $s = 3s_0$. Therefore

$$3s_0 = s_0\alpha X/(\alpha X - X)$$

gives

$$3X(\alpha - 1) = \alpha X$$

leading to

$$3\alpha - 3 = \alpha$$

i.e.

$$2\alpha = 3$$

giving

$$\alpha = 1.5.$$

(ii) If $\alpha = 1.5$, $s_0 = 30$ ppm and $x = X/2$, then

$$s = 45X/(1.5X - 0.5X)$$
$$= 45 \text{ ppm}.$$

(iii)

$$s = s_0\alpha X/(\alpha X - x)$$

gives

$$\frac{ds}{dX} = s_0\alpha X/(\alpha X - x)^2$$
$$= s/(\alpha X - x).$$

(iv) From (ii), $s = 45$ ppm at $x = X/2$. Thus, (iii) above gives the gradient for $X = 10\,000$ m and $\alpha = 1.5$ at $x = X/2$ as

$$\frac{ds}{dX} = \frac{45}{15\,000 - 5000} = 4.5 \times 10^{-3} \text{ ppm m}^{-1}$$

Hence, 1 km either side of centre, salinity changes by $1000 \times 4.5 \times 10^{-3}$ ppm = 4.5 ppm. Thus, 1 km seaward, salinity is 40.5 ppm and 1 km shoreward, salinity is 49.5 ppm.

8.15 (i) $\dfrac{du}{d\theta} = 4\alpha et \sin\theta\cos\theta$

(ii) $\dfrac{dv}{d\theta} = -2(\cos\theta\sin\theta)(\sin^2\theta - \cos^2\theta)$

(iii)
$$\frac{d^2w}{d\theta^2} = \frac{v(du/d\theta) - u(dv/d\theta)}{v^2}$$

$$= \alpha et\,\frac{4(\cos^2\theta\sin^2\theta) + 2(\sin^2\theta - \cos^2\theta)(\sin^2\theta - \cos^2\theta)}{\cos^3\theta\sin^3\theta}$$

$$= \alpha et\left(\frac{4}{\cos\theta\sin\theta} + \frac{2(\sin^2\theta - \cos^2\theta)}{(\cos\theta\sin\theta)^3}\right).$$

8.16 (i) $\dfrac{\mathrm{d}t}{\mathrm{d}x} = \dfrac{-t_0}{X}\exp(-x/X)$, (ii) $-1.098\ \mathrm{m\,km^{-1}} = -0.001\,098$.

CHAPTER 9

9.1 $\displaystyle\int 10\,\mathrm{e}^{2\alpha}\,\mathrm{d}\alpha = 5\,\mathrm{e}^{2\alpha} + k.$

9.2 $y = \dfrac{x^{n+1}}{n+1}$ gives $\dfrac{\mathrm{d}y}{\mathrm{d}x} = \dfrac{1}{n+1}(n+1)x^n = x^n.$

9.3 (i) $\sin(x) + k$, (ii) $\dfrac{\xi^{11}}{11} + k.$

9.4 $\displaystyle\int \frac{10}{x} + \cos(x)\mathrm{d}x = 10\ln(x) + \sin(x) + k.$

9.5 $\displaystyle\int_0^1 4\,\mathrm{e}^x\,\mathrm{d}x = [4\mathrm{e}^x]_0^1 = 4\,\mathrm{e} - 4 = 6.873.$

9.6 (i) See Figure A.13.

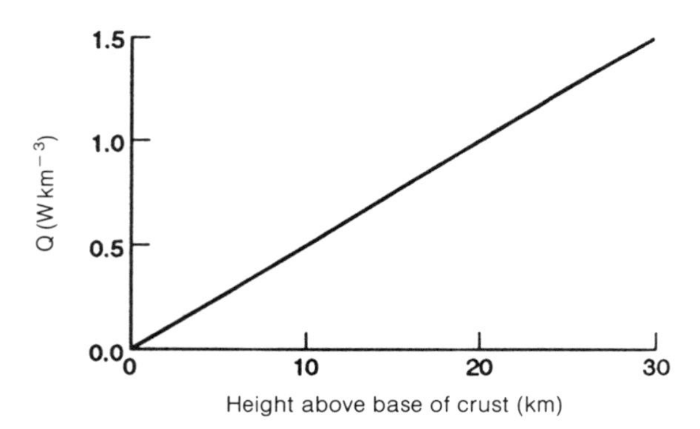

Figure A.13. Question 9.6.

(ii) Rate of heat generation in box $= y\Delta z/20$.

(iii) Approximate total heat generated in column of N boxes is

$$\sum_{i=1}^{N} y_i \Delta z/20.$$

(iv) Total heat generated in column is

$$\int_0^{30} \frac{y}{20}\,\mathrm{d}z$$

(v) Rate of heat flow at surface = $22.5\,\mathrm{kW\,km^{-2}}$.
(vi) $4444\,\mathrm{km^2}$.

9.7 (i) $\pi r^2 \Delta z$.

(ii) $\sum_{i=1}^{N} \pi r_i^2 \Delta z$.

(iii)
$$\int_{-r_p}^{r_p} \pi r^2\,\mathrm{d}z = \int_{-r_p}^{r_p} \pi r_e^2 (1 - (z^2/r_p^2))\,\mathrm{d}z$$
$$= \pi r_e^2 \left(\int_{-r_p}^{r_p} \mathrm{d}z - \frac{1}{r_p^2} \int_{-r_p}^{r_p} z^2\,\mathrm{d}z \right)$$
$$= \pi r_e^2 \left(z - \frac{z^3}{3r_p^2} \right)_{-r_p}^{r_p}$$
$$= \pi r_e^2 \left(r_p - \frac{r_p^3}{3r_p^2} + r_p - \frac{r_p^3}{3r_p^2} \right) = \tfrac{4}{3}\pi r_e^2 r_p.$$

(iv) $1.083 \times 10^{12}\,\mathrm{km^3}$.
(v) $1.081 \times 10^{12}\,\mathrm{km^3}$.

9.8 (i) Area = $t_i \Delta x$.

(ii) Approximate total area = $\sum_{i=1}^{N} t_i \Delta x$.

(iii)
$$\text{Total area} = \int_0^\infty t\,\mathrm{d}x = \int_0^\infty t_0 \exp(-x/X)\,\mathrm{d}x$$
$$= (-Xt_0 \exp(-x/X))_0^\infty = Xt_0.$$

(iv) Total area = $5000\,\mathrm{m^2}$. Thus age = 500 years.

Index

Page numbers in **bold** refer to figures and page numbers in *italics* refer to tables.

Mathematics: A Simple Tool for Geologists

Mathematics: A Simple Tool for Geologists

David Waltham

Department of Geology, Royal Holloway,
University of London, UK

CHAPMAN & HALL
University and Professional Division

London · Glasgow · Weinheim · New York · Tokyo · Melbourne · Madras

Published by Chapman & Hall, 2-6 Boundary Row, London SE1 8HN, UK

Chapman & Hall, 2-6 Boundary Row, London SE1 8HN, UK

Blackie Academic & Professional, Wester Cleddens Road, Bishopbriggs, Glasgow G64 2NZ, UK

Chapman & Hall GmbH, Pappelallee 3, 69469 Weinheim, Germany

Chapman & Hall USA, One Penn Plaza, 41st Floor, New York, NY10119, USA

Chapman & Hall Japan, ITP - Japan, Kyowa Building, 3F, 2-2-1 Hirakawacho, Chiyoda-ku, Tokyo 102, Japan

Chapman & Hall Australia, Thomas Nelson Australia, 102 Dodds Street, South Melbourne, Victoria 3205, Australia

Chapman & Hall India, R. Seshadri, 32 Second Main Road, CIT East, Madras 600 035, India

First edition 1994
Reprinted 1995

Typeset in 10/12 Times by Thomson Press (I) Ltd., New Dehli
Printed in Great Britain at The Alden Press, Oxford

ISBN 0 412 49210 5

A Catalogue record for this book is available from the British Library

Library of Congress Cataloging-in-Publication Data available

∞ Printed on permanent acid-free text paper, manufactured in accordance with ANSI/NISO Z39.48-1992 and ANSI/NISO Z39.48-1984 (Permanence of Paper).